CHEMICAL REACTOR THEORY

Chemical reactor theory

AN INTRODUCTION

K.G. DENBIGH

Director, Council for Science and Society

J.C.R. TURNER

Professor of Chemical Engineering, University of Exeter

THIRD EDITION

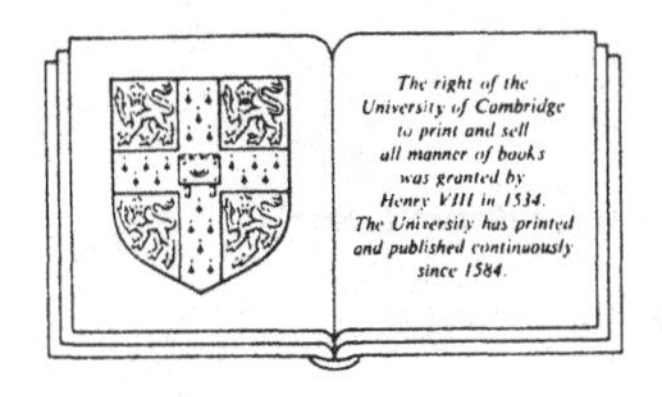

CAMBRIDGE UNIVERSITY PRESS

Cambridge

London New York New Rochelle

Melbourne Sydney

CAMBRIDGE UNIVERSITY PRESS
Cambridge, New York, Melbourne, Madrid, Cape Town,
Singapore, São Paulo, Delhi, Tokyo, Mexico City

Cambridge University Press
The Edinburgh Building, Cambridge CB2 8RU, UK

Published in the United States of America by Cambridge University Press, New York

www.cambridge.org
Information on this title: www.cambridge.org/9780521276306

First published 1965
Reprinted 1966
Second edition 1971
Third edition 1984
Re-issued 2011

A catalogue record for this publication is available from the British Library

Library of Congress Catalogue Card Number: 83-7551

ISBN 978-0-521-25645-2 Hardback
ISBN 978-0-521-27630-6 Paperback

CONTENTS

PREFACE

Since the first edition of this book was written, in 1965, the subject of chemical reaction engineering has greatly expanded, as a result of much research, in universities and in industry. It has become an essential part of most undergraduate teaching, and in fact is a required component of academic courses seeking professional accreditation in the U.K.

The problem is thus how far to go, what to put in, and where to stop. This text has been written primarily for undergraduates, but it is hoped that it will also prove useful to those in industry – perhaps trained as chemists or mechanical engineers – who have not yet made acquaintance with reactor design. There are also those, such as experts in catalysis, whose interest is mainly in molecular processes but who may yet find it useful to know how their studies relate to large-scale chemical production.

The first need of such readers is a sense of orientation within the subject, and a firm grasp of its basic principles. For this reason we have deliberately chosen to emphasize the physically simplest notions. This perhaps carries the risk of oversimplification; most industrial reactors are, after all, more complex than the idealized types we mainly discuss. And again, in an introductory text of this length, little can be said about heat transfer, pressure drop, or control. These subjects, in a chemical engineering course, will usually obtain separate treatment, and it is in the study of reactor design that they can be brought together.

For similar reasons we have not included examples requiring computer solution. Computational skill is now required of all students, but we believe that the most suitable problems for illustrating basic principles are those which can be solved fairly simply once the principles have been understood. With the confidence obtained from such success, modern engineers will have little trouble in using their knowledge in more numerically complicated fields.

In our view first-degree chemical engineering graduates should all have a firm command of much of this book. The material in the first six chapters may be expected to be covered in any such course. The particular interests of the staff available might cause some topics in the last four chapters to be left out, or replaced by fuller treatments. In any case it is hoped that the later chapters will give an introduction to the detailed and intensive work now going on in those areas.

The number of problems in the book has been increased and they are to be found at appropriate places in the text. Many of them closely resemble, or the ideas for them were prompted by, questions set to students at Cambridge, Edinburgh, Exeter, or Imperial College London. We are grateful, as before, for permission to use this material. Teachers using this book can obtain a manual of solutions to the problems by writing to J.C.R.T. at the Department of Chemical Engineering at Exeter University.

We would also like to thank the many friends, in industry and in academic life, who have influenced our thoughts on this subject. They know who they are, and others who do not know should be able to guess from the flavour of the text and the references quoted.

K.G.D. J.C.R.T.
London *Exeter*

1

Introduction: reactor types

1.1 The nature of the problem

The kind of industrial problem this book is concerned with is briefly as follows: how to choose the best type of reactor for any particular chemical reaction; how to estimate its necessary size and how to determine its best operating conditions. When faced with a design problem of this sort, the chemical engineer must usually regard two things as being fixed beforehand: one of them is the scale of operation (i.e. the required daily output) and the other is the kinetics of the given reaction. Apart from these he has considerable freedom of choice: he can adopt either a batch process or one of the several different kinds of continuous process; within limits he can take whatever values he believes best for the initial concentration of the reagents and also for the operating temperatures and pressures; finally, he can make controlled alterations in some of these variables during the course of the reaction. Much of the subject matter of this book is concerned with the logic of these decisions.

1.2 Criteria of choice

In the manufacture of products for sale, the criterion of profitability has usually to be satisfied for the project to go forward. Since no chemical reactor of itself can make a profit, the chemical reaction engineer has to consider the effect of his reactor on the profitability of the project as a whole. There are two main aspects of this issue: (*a*) the cost of the reactor (both capital and running costs); and (*b*) the cost to the rest of the project of working up the product of the reactor to its final, saleable form.

The effect of the project on the society in which it is to be based has also to be considered. On the one hand, the choice of the entire project may be substantially determined by governmental/political factors; the need to provide employment, or to use a local raw material, may, by direction, or by taxation arrangements, decide matters.

Again, society may demand stringent 'environmental standards'. In recent years, certainly in the developed economies, an increasing popular awareness of pollution problems has led to increasingly tight restrictions on the emission of undesirable substances (and most substances come into this category) to the atmosphere or to water courses, while the effect on the environment of the products themselves (e.g. the insecticide DDT) comes under increasing scrutiny.

Finally the safety of the plant, to those working on it as well as to the surrounding populace, is a matter of increasing concern. The examination of a process for potential hazards has become more formalized and rigorous. The choice of process may well depend on the cost of making it safe.

A book such as this must concentrate on the more readily quantifiable aspects of chemical reactor design; matters such as plant yields, construction costs and running costs must therefore predominate. In practice these are only items within a larger picture, and the chemical reaction engineer should not lose sight of that fact.

1.3 Batchwise and continuous reaction

Certain chemicals produced in rather small quantities – pharmaceuticals, dyestuffs and so on – are made batchwise. In a typical factory concerned with this sort of production one may be struck by the presence of numbers of autoclaves – each used for producing a ton of one product one day, and a ton of quite a different product the next. Such a system gives great flexibility, especially when the particular factory has a large repertoire of products each produced on a fairly small scale.

A further advantage of batchwise operation is that the capital cost is often less than for a corresponding continuous process when the desired rate of production is low. For this reason it is frequently favoured for new and untried processes which are to be changed over to continuous operation at a more advanced stage of development, when a larger production may be required.

The reasons why continuous processes are eventually adopted in almost all large-scale chemical industries are mainly these:

(*a*) Diminished labour costs, owing to the elimination of operations, such as the repeated filling and emptying of batch vessels.

(*b*) The facilitation of automatic control. This also reduces labour costs, although it usually requires considerable capital outlay.

(*c*) Greater constancy in reaction conditions and hence greater constancy in the quality of product.

It will be seen that the decision between batchwise and continuous reaction is very dependent on the magnitude of capital costs in relation to operating costs (of which labour costs may be a very important component). What is best for a highly industrialized country is not necessarily best for one which is less industrialized.

Let us ask now what is the scientific, as distinct from the economic, difference between batchwise and continuous reaction. The kinetics of reactions are usually studied in the laboratory under batchwise conditions, but the application of the results to the design of a continuous process involves no new principles of kinetics since the molecular changes are the same. The difference lies in the existence of a state of flow in the continuous process and this may give rise to important changes of a macroscopic kind. In particular not all molecules passing through the flow system will necessarily have equal residence time, nor will these molecules all undergo the same history of concentration or temperature changes. These factors may cause appreciable differences of yield or of mean reaction rate, as compared to a batch process. This is especially the case when the reaction situation is complicated by the existence of competing side reactions. Here the yield of the desired product may differ considerably as between batchwise and continuous operation, and also as between the two main types of continuous process. Reaction yield is not necessarily lower by continuous process – in some instances it may be higher. However, in examples where it is lower, this factor may so outweigh the normal advantages of continuous operation as to argue in favour of a batchwise system.

It may be remarked that certain processes are neither unambiguously batch nor unambiguously continuous, but should be described as 'semi-batch' or 'semi-continuous'. For example, penicillin is made in large fermenters which are inoculated with the penicillin-producing organism at the start of a production run. After many hours, the contents of the fermenter are emptied, and penicillin recovered from them. This would therefore seem to be a batch process. However, during the run air, and nutrients such as sugar, are continuously added to the fermenter, and gaseous waste products are continuously removed.

1.4 The tubular reactor

In considering continuous reaction equipment, it is often convenient to bear in mind certain simple, or idealized, types of reactor. We shall discuss the general characteristics of these 'model' reactors, leaving the more detailed treatment till later. The first type we shall describe is the tubular reactor.

The tubular reactor is so named because in many of its instances it takes the form of a tube. However, what is meant in general by a tubular reactor is any continuously operating reactor in which there is a steady movement of one or all of the reagents in a chosen spatial direction (the reagents entering at one end of the system and leaving at the other)* and in which *no attempt is made to induce mixing* between elements of fluid at different points along the direction of flow: that is to say, it is the type of continuous reactor for which the most appropriate first approximation useful for predicting its behaviour is the assumption that the fluid moves through it like a plug (the description *plug-flow reactor* is frequently used). Some reactors which satisfy this definition and yet which bear no outward resemblance to a tube will be mentioned shortly.

Tubular reactors that actually are tubes are used for many gas reactions and some liquid-phase reactions. As examples we can mention the thermal cracking of hydrocarbons to make ethylene, the oxidation of nitric oxide (one stage in the production of nitric acid from ammonia), and the sulphonation of olefines. For such homogeneous reactions the reactor contains only the reacting fluid.

Tubular reactors are also used extensively for catalytic reactions. Here the reactor is packed with particles of the solid catalyst and for this reason is often referred to as a *fixed-bed reactor.* Such uses include ammonia synthesis, methanol synthesis and a host of other important heterogeneous reactions. The reactor may be one large-diameter cylinder, or may consist of many tubes in parallel, fixed between two headers as in a tube-and-shell heat exchanger. The tubes are frequently a few centimetres in diameter and may be several metres in length. Fig. 1 is a diagram of a 'tube-cooled' ammonia-synthesis reactor. The mixture of nitrogen and hydrogen entering at the top first passes downwards over the internal walls of the steel forging. This is for the purpose of keeping the metal cool. The gas then passes upwards through the nest of tubes and is thereby raised in temperature by heat exchange, firstly with the hot product gases and secondly with the catalyst. In this reactor the catalyst is in the space *outside* the tubes and is supported on a grid. Emerging from the top of the tubes, the gas passed downwards through the catalyst material, with evolution of heat of reaction. It is then partially cooled by the incoming gas and leaves the reactor at the bottom.

* In certain instances, however, there may be advantages in introducing subsidiary streams of one or other of the reagents along the path of the reaction.

The diagram also shows bypass and heating arrangements which are necessary at 'start-up' and 'shut-down' of the reactor. In the design of any reactor attention must be paid not only to the performance of the reactor when running steadily, but also to the means of arriving at that state and of stopping the process, either for routine maintenance or in the case of breakdown.

In the above reactors the reactants flow axially down a tube, but there are others where the flow is not so simple. In a design of reactor which has been applied to the 'platforming' reaction, the reactant gas flows

Fig. 1. Tube-cooled ammonia-synthesis reactor. (*a*) Gas inlet to sheath and interchanger, (*b*) gas inlet bypassing interchanger, (*c*) gas inlet bypassing tubes, (*d*) gas exit, (*e*) pyrometer, (*f*) heater electrode. (By courtesy of Imperial Chemical Industries PLC.)

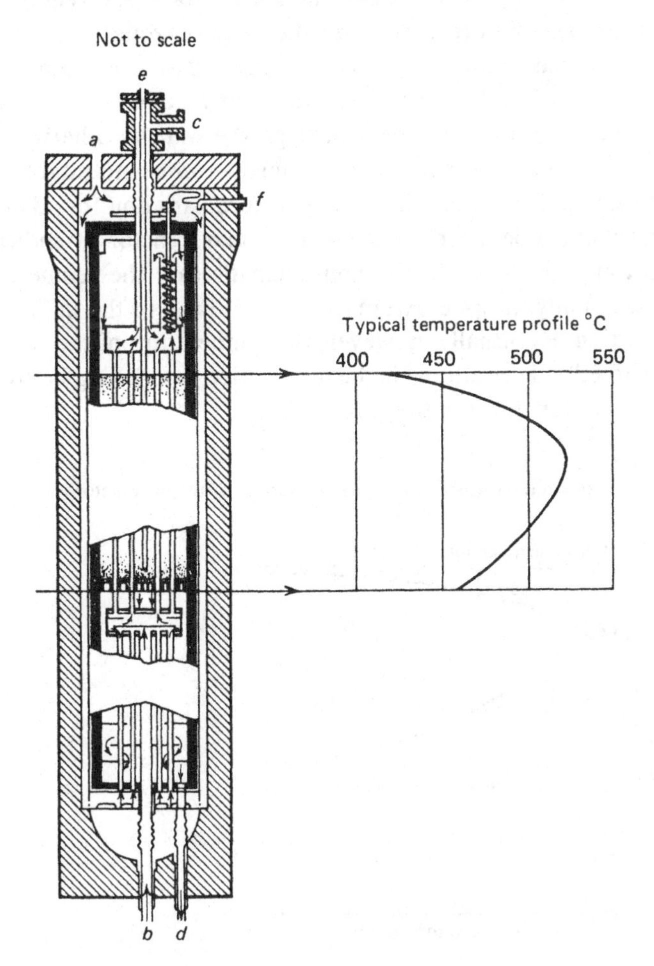

radially inwards from a perforated cylinder through a catalyst packed between this cylinder and a central perforated tube which collects the product. Though the gas flow is radial, the system satisfies the 'plug-flow' requirement in so far as no attempt is made to cause mixing between points in the direction of flow.

In all such reactors the composition of the reacting fluid necessarily changes in the direction of flow. However, there may also be variations of composition in directions at right-angles to the direction of flow and such variations can be the result of temperature gradients and/or velocity gradients.

Tubular reactors are sometimes operated adiabatically and sometimes with heat transfer through the wall. In the former case the temperature naturally rises along the direction of flow if the reaction is exothermic and falls if it is endothermic. Of course, in many instances, it is necessary to heat the reactants before they enter the reaction zone (e.g. by heat exchange, as in the ammonia-synthesis reactor discussed above), as otherwise reaction would be too slow. But once reaction is under way, it is often necessary to remove heat through the wall, as otherwise the temperature rise might be excessive and might cause undesirable side reactions to set in. The longitudinal temperature variation for adiabatic and non-adiabatic operation of such a tubular reactor is indicated diagrammatically in Fig. 2. In the non-adiabatic case the temperature tends to rise initially, despite external cooling, owing to the initial high speed of reaction. Eventually, however, the temperature of the reacting fluid begins to fall, the result of the heat evolution rate falling below the rate of heat transfer through the wall.

Fig. 2. Longitudinal variation of temperature in a tubular reactor.

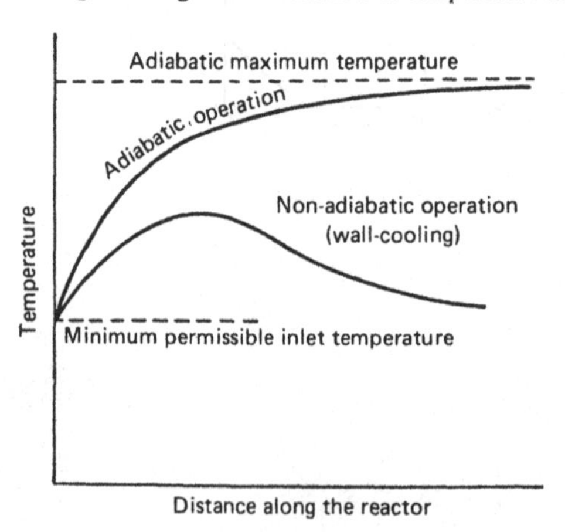

Transverse (or radial) variations of temperature also tend to occur, especially if the reacting fluid and the catalyst (if present) have low thermal conductivities. With exothermic reactions the highest temperatures naturally occur in those parts of the reactor furthest from the surface where the heat is taken out. This can be a very considerable effect. Fig. 3 shows the temperature contours occurring in a catalyst bed where SO_2 was being oxidized [**4**]. The wall of the reactor was maintained at 197 °C; temperatures as high as 520 °C were found in the interior of the catalyst, although the diameter of the cylindrical reactor was only 5 cm.

The occurrence of a 'hot spot' such as is shown in the figure, can have very adverse effects. For example in methanol synthesis,

$$CO + 2H_2 = CH_3OH,$$

as carried out at elevated pressures, any region in the catalyst where the

Fig. 3. Temperature contours in a tubular reactor. (Based on J. M. Smith, *Chemical Engineering Kinetics*, McGraw-Hill.)

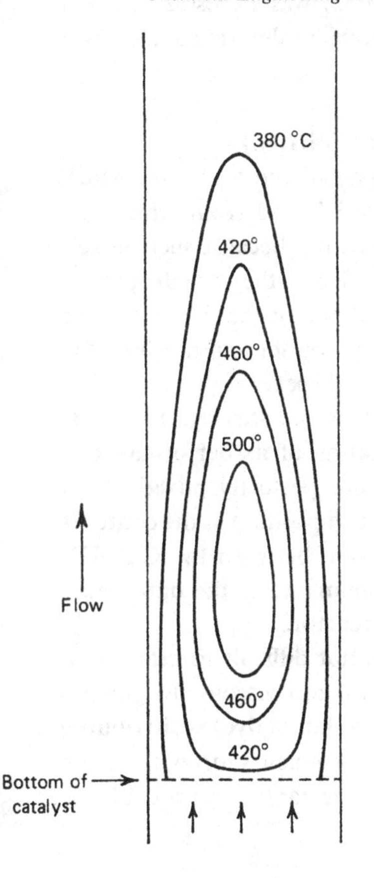

temperature is excessive can give rise to the onset of an undesired reaction, the formation of methane.

For efficient removal of heat, the diameter of a cylindrical tubular reactor should clearly be small, in order to reduce the distance over which the heat must be conducted up to the wall. When there are compelling reasons of a different kind for choosing a large diameter, or for placing the catalyst in large trays, it may be necessary to immerse cooling coils in the body of the catalyst.

A striking type of reactor having mercury cooling is that which was used in a process for the production of phthalic anhydride. Naphthalene was vaporized into an air stream and passed through a tubular reactor consisting of as many as 3000 tubes in parallel, each 1–2 cm in diameter and up to 3 m long and containing the pelleted catalyst. The construction was similar to that of a tube-and-shell heat exchanger. The heat of reaction was very effectively taken up by the boiling of mercury outside the tubes; the mercury vapour was condensed externally and recirculated. Careful temperature control, at about 350 °C, was necessary in this reaction in order to reduce the formation of maleic anhydride and carbon dioxide as by-products.

1.5 The continuous stirred tank reactor (C.S.T.R.)

This type of reactor consists of a well-stirred tank into which there is a continuous flow of reacting material, and from which the (partially) reacted material passes continuously. It is because such vessels are squat in shape (e.g. cylindrical vessels as wide as they are deep) that good stirring of their contents is essential; otherwise there could occur a bulk streaming of the fluid between inlet and outlet and much of the volume of the vessels would be essentially dead space.

The important characteristic of a C.S.T.R. is the stirring. The most appropriate first approximation to an estimation of its performance is based on the assumption that its contents are perfectly mixed. As a consequence the effluent stream has the same composition as the contents and this demonstrates the important distinction between the C.S.T.R. and the tubular reactor. Fig. 4 shows diagrammatically the differences between four commonly occurring types of reactor.

A fair approximation to perfect mixing is not difficult to attain in a C.S.T.R., provided that the fluid phase is not too viscous. In general terms, if an entering element of material (e.g. a shot of dye) is distributed uniformly throughout the tank in a time very much shorter than the average time of residence in the tank, then the tank can probably be taken to be 'well-mixed'.

It will be seen later that it is often advantageous to have several C.S.T.R.s in series, the process stream flowing from one to the next. This results in a *stepwise* change of composition between successive tanks. It also results in a *bypassing loss*: a given molecule entering a tank has a significant probability of finding its way into the outflow almost immediately. This is the reason why it is usually necessary to use several tanks in series, especially if high conversions are required. If there were only one or two there would be an appreciable loss of unreacted reagent. Although this loss is, in a sense, the result of the stirring, the loss would usually be even greater in the absence of stirring, due to bulk streaming between inlet and outlet.

A usual consequence of the stepwise change in concentration is that the average reaction rate is lower than it would be in a tubular reactor having the same concentrations of reagents in the feed.* The reactor volume, for a given output, must therefore be a good deal larger and this must be allowed for in the design. In view of this it may seem paradoxical that the C.S.T.R. is much used for comparatively slow liquid-phase reactions. The reason is the cheapness of the construction, as compared with a tubular reactor. The greater volume is comparatively unimportant as an economic factor, at any rate in the case of tanks

Fig. 4. Main types of reactor.

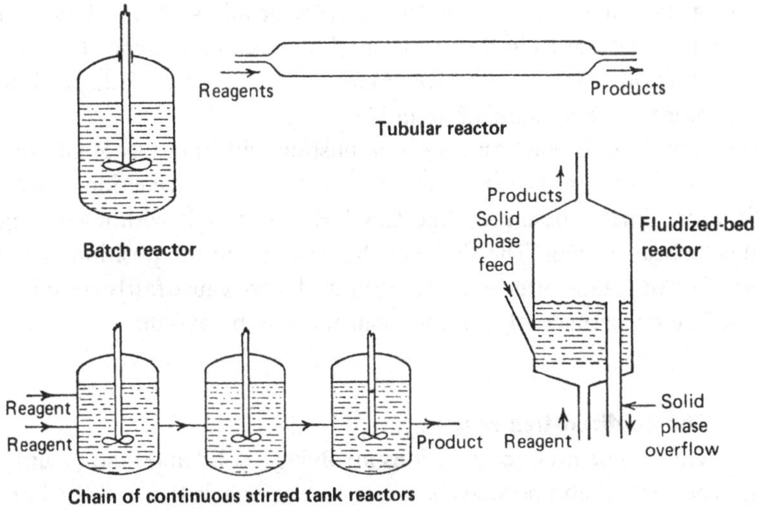

* However, certain exceptional types of kinetics, and especially autocatalysis, may result in a *higher* rate in a C.S.T.R. than in a tubular reactor. This occurs in certain biochemical processes.

operating at atmospheric pressure and made of an inexpensive material such as mild steel.

One great advantage of the C.S.T.R., apart from simplicity of construction, is the ease of temperature control. The reagents entering the first vessel plunge immediately into a large volume of partially reacted fluid and, because of the stirring, local hot spots do not tend to occur. Also the tanks of the C.S.T.R. offer the opportunity of providing a very large area of cooling surface. In addition to the external surface of the vessels themselves, a large amount of internal surface, in the form of submerged cooling coils, can be provided. Sometimes, in place of coils, a calandria is used, as in the example of the Schmid nitrator for nitroglycerine.

A further advantage, as compared to the tubular reactor, is the openness of the construction. This makes it easy to clean the internal surfaces and this is important in the case of reactions where there is a tendency for solid matter to be deposited, e.g. polymerization processes and reactions in which tarry material is formed as a by-product.

For these various reasons the typical fields of application of the C.S.T.R. are continuous processes of sulphonation, nitration, polymerization, etc. It is used very extensively in the organic chemical industry and particularly in the production of plastics, explosives, synthetic rubber and so on. The C.S.T.R. is also used whenever there is a special necessity for stirring; for example, in order to maintain gas bubbles or solid particles in suspension in a liquid phase, or to maintain droplets of one liquid in suspension in another as in the nitration of benzene or toluene. The rate of such reactions can be very dependent on the degree of dispersion, and therefore on the vigour of agitation.

Certain gas-phase reactions, e.g. combustion, chlorination of gaseous hydrocarbons, etc., are also sometimes carried out in reaction vessels which approximate to single-stage C.S.T.R.s, although without having any mechanical stirring. The shape of the vessels and the position of the gas inlet jets (e.g. tangential entry) may be such as to cause fairly complete mixing. The occurrence of hot spots can thereby be avoided.

1.6 The fluidized-bed reactor

This can be used for reactions involving a solid and fluid (usually a gas). The earliest and perhaps best-known example is the 'cat-cracker', in which the solid is a catalyst for the cracking of hydrocarbon vapours. A similar catalytic process has been used for the oxidation of naphthalene with air to give phthalic anhydride.

Examples of processes in which the solid actually reacts with the gas are the reaction of alumina with hydrogen fluoride to produce aluminium fluoride, and the reactions

$$UO_3 \xrightarrow{H_2} UO_2 \xrightarrow{HF} UF_4,$$

which are carried out in fluidized-bed reactors.

In such reactors the solid material in the form of fine particles is contained in a vertical cylindrical vessel. The gas stream is passed upwards through the particles at a rate great enough for them to lift, but not so great that they are prevented from falling back into the fluidized phase above its free surface by carry-over in the gas stream. The bed of particles in this condition presents the appearance of boiling; bubbles of the upflowing gas may be seen bursting at the upper surface.

The effect of the rapid motion of the particles is a high degree of uniformity of temperature and thus an avoidance of the hot spots which occur with fixed-bed tubular reactors. This is often a considerable advantage in the case of reactions which can be allowed to proceed adiabatically, finding their own temperature as determined by the heat of reaction. If heat must be removed, in order to keep the temperature down to some prescribed level, use can be made of the fact that heat transfer to cooling tubes is more easily brought about in a fluidized bed than in a fixed bed. Care must be taken to ensure that the heat-transfer surface does not interfere with the efficiency of fluidization. A disadvantage of this type of reactor is attrition of the solid, resulting in, for example, loss of catalyst and dust problems in the effluent gas stream.

What fluidization offers is fully continuous operation under conditions where otherwise it might be impossible. This may be illustrated by reference to catalytic cracking. This particular reaction is accompanied by the rapid deposition of carbon on the surface of the catalyst. Under the conditions of fluidization, the solid phase can be continuously withdrawn from the reactor as if it were a liquid (Fig. 4) and circulated to a fluidized regenerator, where the carbon is burnt off in an air stream, and thence returned to the cracker. More generally it will be seen that the use of a circulating catalyst which is continuously regenerated is the logical completion of the trend towards fully continuous processes. The system becomes continuous in operation with respect to the *catalyst*, as well as the reagents, and the effect is therefore to eliminate the need for stand-by reactors such as are necessary in fixed-bed processes, to allow for periods of catalyst replacement, or regeneration.

The fluidized reactor is difficult to treat theoretically because of the complicated nature of the fluid flow. With gas-fluidized beds much of the gas may pass through the bed as bubbles. The reaction only occurs on the surface of the solid particles and the flow of bubbles through the bed acts as a bypass stream. Gaseous reactant can move from the bubble phase to the particulate phase by diffusion and by convection (there is a convective flow *through* the bubbles, which do not have 'skins'). The relative importance of reaction, diffusion and convection depends upon the fluid mechanics, which are not precisely known.

It is particularly difficult to scale up fluidized-bed reactors from laboratory experiments. The behaviour of bubbles in relatively shallow fluidized beds in small-diameter tubes can be very different from their behaviour in deep, large-diameter beds. For example, slugging behaviour may occur in narrow tubes, and bubble coalescence may be important in deep beds.

1.7 Other types of reactor

There are many other, diverse, types of reactor which have been found useful in particular cases. Brief mention can be made of four.

(i) Bubble-phase reactors

Here a reagent gas is bubbled through a liquid with which it can react, because the liquid contains either a dissolved involatile catalyst or another reagent. The product may be removed from the reactor in the gas stream. Mass transfer is clearly important, and may control the rate of 'reaction'. An example is the Hoechst–Wacker process for the production of acetaldehyde by the oxidation of ethylene.

(ii) Slurry-phase reactors

These are similar to bubble-phase reactors, but the 'liquid' phase consists of a slurry of liquid and fine solid catalyst particles. They are used in the catalytic carbonylation of hydrocarbons to produce alcohols. 'Three-phase fluidized beds' are of this type.

(iii) Trickle-bed reactors

In these reactors the solid catalyst is present, not as fine fluidized particles, but as a fixed bed. The reagents, which may be two partially-miscible fluids, are passed either co-currently or counter-currently through the bed. An example is the high-pressure hydration of propylene to give isopropyl alcohol.

(iv) Moving-burden bed reactors

Here a fluid phase passes upwards through a packed bed of solid. Solid is fed to the top of the bed, moves down the column in a closely plug-flow manner, and is removed from the bottom. This has been used in the catalytic isomerization of xylenes, and in continuous water treatment by ion-exchange.

1.8 The steady state

In batchwise reaction the change of composition occurs in the time co-ordinate. Whether or not a batch system is uniform throughout its spatial co-ordinates, it always changes from moment to moment, and does so for as long as is needed for thermodynamic equilibrium to be attained (or until the process is brought to an end). The significant difference in continuous reaction is that the corresponding change of composition is shifted to a space co-ordinate; any part of the system usually tends towards a time-invariant state and the variation of composition shows itself between one region of the system and another, e.g. between adjacent tanks of a C.S.T.R. or between neighbouring cross-sections of a tubular reactor.

The statement that such a system usually tends towards a time-invariant state applies, of course, only if there are constant feed conditions, constant rates of heat removal, etc. Even so, a reaction system does not of necessity approach a steady state, and in special cases it may occur that the concentrations of the various substances which are present will oscillate continuously about particular values. This can occur in complex autocatalytic reactions, including enzyme reactions and also reactions having special thermal characteristics.

It is important to note that the steady state, if it is attained, is not an *equilibrium*. This term should be reserved for the time-independent state of closed systems. The steady state of an open system, such as a continuous reactor, depends on the flow regime, the reaction rates, and the size of the system. An economic balance has usually to be struck between the value of an increased yield of product and the cost of the larger reactor necessary to produce it.

1.9 Transient behaviour

This book will be mainly concerned with continuously reacting systems when they have already attained the steady state. However, for certain purposes it is important to know how these systems behave when they are not in the steady state. Some typical applications of the study of such transient behaviour are: (1) the calculation of the time required for the system to approach to within, say, 1% of the steady state and the determination of the start-up procedure which will reduce this time to a minimum; (2) investigation of the type of product obtained during the approach to the steady state; (3) calculation of the speed at which changes or fluctuations at the reactor inlet are transmitted to the outlet, or to any intermediate point in the system.

There is a considerable literature dealing with the stability and the automatic control of continuous reactors. The increasing availability, and decreasing real cost, of computers and microprocessors are leading to a widespread increase in their use for 'on-line' process analysis and control.

1.10 Factors affecting performance

The products of a chemical reactor will depend upon the reactions which the reagents in the feed may undergo. We require to know what these reactions are, and how their rates depend upon the concentrations and temperature.

It is thus necessary to know how long any element of the material passing through the reactor stays in the reactor, and what is its environment (in respect of concentration and temperature) while it is inside.

We shall, in the course of the following chapters, consider those aspects of chemical kinetics which are of concern to chemical engineers. We shall then be able to consider the different types of reactor in more detail than has been done in this introduction.

References

Recently several new textbooks have been produced, to join more recent editions of older texts. The following is a selection.

1. Aris, R., *Elementary Chemical Reactor Analysis* (Prentice-Hall, 1969).
2. Levenspiel, O., *Chemical Reaction Engineering* (Wiley, 1972).
   To this must be added *The Chemical Reactor Omnibook*, and *The Chemical Reactor Minibook*, (Oregon State University Book Stores Inc., Corvallis, 1979).
3. Carberry, J. J., *Chemical and Catalytic Reaction Engineering* (McGraw-Hill, 1976).
4. Smith, J. M., *Chemical Engineering Kinetics* (3rd edn, McGraw-Hill, 1981).
5. Hill, C. G., *An Introduction to Chemical Engineering Kinetics and Reactor Design* (Wiley, 1977).
6. Holland, F. A. and Anthony, R. G., *Fundamentals of Chemical Reaction Engineering* (Prentice-Hall, 1979).
   As well as the above texts the proceedings of the various International Symposia on Chemical Reaction Engineering are published by Pergamon Press, and these report many of the advances at the research level.

2

Chemical kinetics

2.1 Introduction

Although this is not a book on chemical kinetics as such, it is convenient to give a brief review of those aspects of the subject which should be known to the chemical engineer when he is considering reactor design and performance.

One may discern three levels of understanding of a chemical process. The lowest is one at which we can say that if certain operating conditions are maintained a certain output can be produced. At the next level we might be able to say that the various reactions obey a given set of equations, with more or less well-defined rate constants. These constitute a mathematical model which can be used by the chemical engineer to design a plant or to control the performance of an already existing plant.

At the most sophisticated level we can say why the rate constants have the values they do have and why certain reaction paths are favoured. We can interpret the system in terms of molecular mechanisms. Broadly speaking this level is not the concern of the chemical engineer and process designer. It *is* the concern of the research chemist in industry, who may be led by such considerations to the discovery of entirely new processes.

As chemical engineers, for the purposes of this book, we shall be interested in formulating mathematical models. These should, of course, be in quantitative agreement with experimental information. We are not necessarily concerned with explaining why reactions follow certain paths.

The two basic equations required to describe reactor performance are the material balance equation and the rate equation. A clear understanding of these simple equations and a willingness to write them down are essential. The authors have seen generations of students get into difficulties as a result of neglecting these first steps.

2.2 The material balance equation

Consider a small volume element of a reactor and what happens there in a small interval of time. A material balance on any reactant species can be drawn up. It is as follows:

Moles entering element (1) =

Moles leaving element (2) + Moles reacting (3) + Change of moles within element (4) (2.1)

These four terms comprise the material balance. In a particular case, any one of the four terms may be zero. For instance, in a batch reactor terms (1) and (2) will be zero. In a physical separation process, term (3) will be zero. Again, for a reactor *in the steady state* term (4) will be zero. Each of these constitutes a special case; the student should always think first of equation (2.1) and not try to short-cut to the special case. Terms (3) and (4) may be either negative or positive. To put the point more precisely, if the chosen species is a *product* of the reaction, term (3) should be written '−Moles formed'.

2.3 The rate equation

To be able to calculate term (3) we must know the rate equation for the particular species. If this species is a reactant, its rate of reaction may depend upon the concentrations of several species, including its own, and it is customary to speak of a reaction as being 'nth order', where n is commonly integral and usually not greater than 3. We must now be more precise in our definition. In saying that a given reaction is nth order *with respect to a given species, A*, we mean:

{Moles reacting per unit volume per unit time in a given volume element at a given time}

is proportional to {the concentration of A in that element at that time} raised to the nth power. (2.2)

Consider an example. We are told that 'the decomposition of A is first order with respect to A', and we wish to formulate the rate expression. A common response of the student, amounting almost to a reflex action, is to write down

$$\frac{dc_A}{dt} = -kc_A, \tag{2.3}$$

where k is the first-order velocity constant and c_A is the concentration of A at time t. This equation is *not* in general correct. It is correct *only* for the case of a constant-volume batch reactor, in which case it is equations (2.1) and (2.2) combined. Because chemists use constant-volume batch reactors for many, if not most, of their laboratory experiments, and because student chemical engineers are often taught chemistry by chemists, it is all too commonly believed by students that equation (2.3) defines what is meant by a first-order reaction. The correct equation can be written

$$r = kc_A. \tag{2.4}$$

This can then be substituted in equation (2.1), bearing in mind that the rate, r, in equation (2.4) means moles of A reacting per unit volume per unit time.

***Example* 2.1**

Show that equation (2.3) is correct for the first-order decomposition of A in a constant-volume batch reactor.

Solution

Consider equation (2.1). Since we have a batch reactor, terms (1) and (2) are both zero.

From equation (2.4) the amount of A reacting in time dt in an element of volume dV is $kc_A \cdot dV \cdot dt$, and this is term (3) in equation (2.1). For term (4), the change of moles within the element, we have $d(c_A\, dV)/dt \cdot dt$. Equation (2.1) therefore becomes

$$\frac{d}{dt}(c_A\, dV) = -kc_A\, dV. \tag{2.5}$$

We assume that our batch reactor is uniform in temperature and pressure. We can therefore integrate over the whole volume at a given time, taking c_A and k to be constant. We obtain

$$\frac{d}{dt}(c_A V) = -kc_A V. \tag{2.6}$$

Since this is a *constant-volume* batch reactor, equation (2.6) becomes

$$\frac{dc_A}{dt} = -kc_A,$$

which is equation (2.3). If the reactor is isothermal as well, then k is independent of time and (2.3) is easily integrated.

We have stressed the logic behind equations (2.1) and (2.2) because the reader should seek to accustom himself to thinking along these lines. The simple expressions for particular cases of equation (2.1) will be obtainable when required, and the less simple situations will be amenable to this treatment as well. The reader should work through the following example, using the logic of Example 2.1.

***Example* 2.2**

A gas, A, decomposes irreversibly to form a gas, B, according to the reaction

$A \rightarrow 2B.$

This reaction is known to be first order with respect to A. The decomposition is carried out in an isothermal, constant-pressure, batch reactor. Derive the differential equation which governs the volume of the system as a function of time if the reacting gas may be assumed to behave as a perfect gas mixture.

[*Answer.* $dV/dt = k(2V_0 - V)$, V_0 = initial volume.]

The application of equations (2.1) and (2.2) to different situations will be demonstrated in later chapters dealing with reactor types in more detail.

2.4 Reaction order and velocity constants

Generalizing what has been said already, in simple gas reactions it is known from experiment as well as from theory that the reaction rate is proportional to the product of small integral powers, α and β, of the concentrations of the reagents. Thus

$$r = k[A]^\alpha[B]^\beta, \tag{2.7}$$

where k is a proportionality factor, the *velocity constant*, which depends only on temperature. That such an expression should hold is to be expected from the kinetic theory of gases, since molecular collision rates are determined by the products of molecular concentrations.

The powers α and β define the order of the particular reaction. For example, the rate of oxidation of nitric oxide is proportional to the product $[NO]^2[O_2]$, and the reaction may be described as being either third order, or, more explicitly, as being second order with respect to nitric oxide and first order with respect to oxygen. The powers α and β are usually 0, 1 or 2, but simple fractional powers such as $\frac{1}{2}$ may also occur. It is to be noted that the order of a reaction is determined empirically, by experiment, and it is subsequently for theory to decide why the particular reaction has this particular order. This involves the setting up of a reaction *mechanism* and this may be much more complex than the simple stoichiometry of the reaction might seem to imply. For example, the occurrence of the order $\frac{1}{2}$ may often be taken as signifying that the mechanism involves a dissociation into free atoms or radicals.

It is important to notice that the numerical value of the velocity constant depends on which of the reactants or products is chosen as the key substance for the purpose of defining the reaction rate. For example, if the stoichiometry of the reaction happens to be

$$A+B=2C,$$

the relationship between the rates of change of mole numbers in a batch reactor is

$$\frac{dn_c}{dt}=-\frac{2dn_a}{dt}=-\frac{2dn_b}{dt}.$$

The number of moles of C increases twice as fast as the number of moles of A or of B diminishes, and it follows that a velocity constant defined with respect to the decrease of A (or of B) would have only half the value of that defined with respect to the increase of C. In using the velocity constants given in the literature it is therefore essential to keep in mind the particular key substance chosen by the authors. There is less risk of ambiguity if the reaction rate is defined by

$$r=\frac{1}{\nu_i V}\frac{dn_i}{dt}, \tag{2.8}$$

where ν_i is the stoichiometric number of the substance i taking part in the reaction (ν_i is positive for products, negative for reactants). A rate defined in this way has the same value for all substances involved in the reaction. Another, equivalent, way of resolving the ambiguity is to work in terms of the 'extent of reaction', ξ, which for a constant-volume batch reactor is readily related to the concentration by the equation

$$c_i=c_i^0+\nu_i\xi, \tag{2.9}$$

where c_i^0 is the initial concentration of species i. In this case the rate r can then be unambiguously defined by

$$r=\frac{d\xi}{dt}, \tag{2.10}$$

and this principle can be applied to situations other than the constant-volume batch reactor.

The practice of using equations such as (2.8) or (2.10) is spreading, but it remains true that the value of any velocity constant should still, for clarity, be accompanied by the rate equation to which it refers.

If the reaction is 'reversible', i.e. at equilibrium the conversion to products is appreciably incomplete, then equation (2.7) is inadequate. For such reactions a common form is

$$r=k[A]^{\alpha}[B]^{\beta}-k'[C]^{\gamma'}, \tag{2.11}$$

where γ' is again usually a simple power. The first term represents the rate of the forward reaction, while the second gives the rate of the backward reaction (with its velocity constant k'). Both these processes, e.g. $A+B \rightarrow C$ and $C \rightarrow A+B$, occur simultaneously at the molecular level. At equilibrium, the net rate of reaction r is zero, but both forward and backward reactions may be proceeding rapidly, though at equal rates.

2.5 Thermodynamic restrictions on rate equations

So far we have paid little attention to the stoichiometry of the reaction, and we have not considered any thermodynamic restrictions there may be on the rate equation. In the case of many reactions the powers α, β and γ' are the same as the *stoichiometric coefficients* occurring in the chemical equation of the reaction. Such, for example, is the situation in the reaction

$$2HI = H_2 + I_2,$$

where it is known from careful experiments [**1**] that the reaction rate can be represented very accurately by the equation

$$r = k[HI]^2 - k'[H_2][I_2].$$

However, a very different situation occurs in the formation of phosgene. Here the chemical equation is

$$CO + Cl_2 = COCl_2,$$

but experiments show that the reaction rate is given by

$$r = k[CO][Cl_2]^{3/2} - k'[COCl_2][Cl_2]^{1/2}.$$

In the first of these two reactions, the powers α, β and γ' correspond to the stoichiometric coefficients; in the second they do not. The choice of α, β, etc. is not unrestricted, however; they must be such as are consistent with thermodynamics when the reaction comes to equilibrium. A more general form of equation (2.11) would be

$$r = k[A]^{\alpha}[B]^{\beta}[C]^{\gamma} - k'[A]^{\alpha'}[B]^{\beta'}[C]^{\gamma'}, \tag{2.12}$$

which allows for the possibility that the products of the reaction may influence (e.g. 'autocatalyse') the rate of the forward reaction and that the reagents may similarly influence the rate of the reverse reaction (see the phosgene reaction above). At equilibrium, when $r = 0$, equation (2.12) reduces to

$$\frac{[C]_e^{\gamma'-\gamma}}{[A]_e^{\alpha-\alpha'}[B]_e^{\beta-\beta'}} = \frac{k}{k'} \tag{2.13}$$

where the subscript e denotes concentrations which are no longer arbitrary but which satisfy the condition of equilibrium. Let the stoichiometric equation, now to be introduced into the discussion for the first time, be

$$aA+bB=cC,$$

where a, b and c are the stoichiometric coefficients. Let it be supposed, for simplicity, that the system in question is a perfect gas mixture. Therefore, we have from thermodynamics

$$\frac{[C]_e^c}{[A]_e^a[B]_e^b}=K, \tag{2.14}$$

where K is a quantity which depends only on temperature and is known as the *equilibrium constant.* Since the ratio k/k' on the r.h.s. of equation (2.13) is also completely determined by the temperature, it follows that this ratio can be regarded as a function of K.

The various powers occurring on the concentrations must now be such as to make equations (2.13) and (2.14) mutually compatible. This will be so if

$$\frac{k}{k'}=K^n, \tag{2.15}$$

which is to say if

$$\frac{[C]_e^{\gamma'-\gamma}}{[A]_e^{\alpha-\alpha'}[B]_e^{\beta-\beta'}}=\frac{[C]_e^{cn}}{[A]_e^{an}[B]_e^{bn}}. \tag{2.16}$$

Here the only restriction on n is that it must be a positive number. (A negative one would imply that the reaction, when the system is *not* at equilibrium, could occur in the direction of free energy increase and this is impossible.) It follows from equation (2.16) that $\gamma'-\gamma$, $\alpha-\alpha'$ and $\beta-\beta'$ *must each be the same multiple n of the stoichiometric coefficients a, b and c.* Thus

$$\frac{\alpha-\alpha'}{a}=\frac{\beta-\beta'}{b}=\frac{\gamma'-\gamma}{c}=n. \tag{2.17}$$

In the example of the phosgene reaction, already discussed, $n=1$, but this is not always the case. For instance, in the formation of nitric acid the stoichiometry can be represented by the equation

$$3N_2O_4+2H_2O=4HNO_3+2NO,$$

but the reaction rate, at constant acid concentration, is given by the following function of the gas composition [**2**]:

$$r=k[N_2O_4]-k'[N_2O_4]^{1/4}[NO]^{1/2}.$$

In this instance thermodynamic consistency is obtained with $n=\frac{1}{4}$. In general it may be said that whenever $n\neq 1$ this arises through having chosen the stoichiometric equation, and thereby the form of the equilibrium constant, in a manner which is not immediately related to the kinetics of the reaction. In the above instance the stoichiometry could equally well be represented by the chemical equation

$$\tfrac{3}{4}N_2O_4+\tfrac{1}{2}H_2O=HNO_3+\tfrac{1}{2}NO.$$

With this equation the equilibrium constant would be the fourth root of that represented by the previous choice of chemical equation and we should now have $n=1$. The value of n is that which connects our choice of writing the stoichiometric equation (which is arbitrary with respect to a numerical multiplier) with the kinetic equation (which reflects the chemical mechanism, and is not a matter of choice).

So much for the situation concerning perfect gas mixtures where volume concentrations can be used satisfactorily for expressing the true thermodynamic equilibrium constant as well as the rate. The situation concerning reactions in solutions or in non-perfect gases is much more complicated and can only be mentioned very briefly. Here the true equilibrium constant has to be expressed in terms of activities – or alternatively concentrations multiplied by activity coefficients. For this reason it was one time suggested that for, for example, a reaction

$$A+B=2C,$$

the rate should be expressed in the form

$$r=ka_Aa_B-k'a_C^2,$$

or alternatively (which comes to the same thing)

$$r=k[A][B]\gamma_A\gamma_B-k'[C]^2\gamma_C^2, \tag{2.18}$$

where the a's are activities and the γ's are activity coefficients. Such expressions would, of course, give complete consistency with the thermodynamic requirements at equilibrium. Later it was realized that expressions of the above form were unnecessarily restricted and that the thermodynamic requirements could be satisfied equally well if each term on the r.h.s. of equation (2.18) were multiplied by the same factor ϕ, which may itself be any function whatsoever of the concentrations. Thus

$$r=k[A][B]\gamma_A\gamma_B\phi-k'[C]^2\gamma_C^2\phi \tag{2.19}$$

and ϕ satisfactorily cancels at equilibrium.

It has then to be decided what value should be ascribed to ϕ. The accurate experimental observations on the decomposition of HI have been examined to this end [**3**, **4**]. It has been shown that, for this reaction, equation (2.18), i.e. $\phi=1$, gives *less* good agreement with results over a range of pressures than the simple equation

$$r=k[A][B]-k'[C]^2. \tag{2.20}$$

Agreement *better* than is obtainable from equation (2.20) may be achieved by using the transition state theory and by assuming the rate to depend upon the concentration of the activated complex. Then ϕ is equal to the reciprocal of the activity coefficient of the activated complex [**3**].

Unfortunately this procedure, which requires a knowledge of ϕ, as well as of the activity coefficients of reagents and products, cannot usually be carried out in practical instances. For this reason it is customary to *define* rate constants by the use of equations such as (2.20). It must then be understood that such 'constants' may be somewhat dependent on the concentrations and it must therefore be carefully noted, by reference to the original experimental work, over what range of the concentrations the k's are sufficiently constant for the purposes in hand.

Turning now to the influence of temperature, it is known experimentally that if the logarithm of the velocity constant is plotted against the reciprocal of the absolute temperature, an almost straight line is usually obtained. (When it is not, this is an indication that the reaction mechanism is not simple and may change as the temperature is changed.) An 'activation energy' E is defined by the relation

$$\frac{d \ln k}{d(1/T)} = -\frac{E}{R}, \quad \text{or} \quad \frac{d \ln k}{dT} = \frac{E}{RT^2} \tag{2.21}$$

where R is the gas constant and T is the absolute temperature. Since an almost straight line plot implies an almost constant value of E, equation (2.21) can be approximately integrated to give

$$k = Z e^{-E/RT}, \tag{2.22}$$

where Z is an integration constant. This is the Arrhenius equation and it is consistent with molecular theories of chemical kinetics, such as the collision theory and the transition state theory. According to the simple collision theory the quantity Z, known as the 'frequency factor', is proportional to the collision rate of molecules at unit concentrations* whilst the Boltzmann factor, $e^{E/RT}$, represents that fraction of these collisions which are energetic enough to lead reaction. The energy of activation E is commonly in the range 10 000–50 000 calories per gram mole, or in SI units, 4×10^7–2×10^8 J kmol^{-1}.

Consider a reversible reaction having velocity constants k and k' which are associated with activation energies E and E' respectively. It will be supposed for simplicity that the reaction occurs in a perfect gas mixture. From equation (2.15):

$$\ln k - \ln k' = n \ln K,$$

*This is not to say, however, that Z is the same for different reactions. Z is regarded as being the product of the collision rate and a 'steric factor' whose value varies widely.

where n is a positive number. Since K and the velocity constants depend only temperature,

$$\frac{d\ln k}{dT}-\frac{d\ln k'}{dT}=\frac{nd\ln K}{dT}.$$

Hence from equation (2.21):

$$\frac{E-E'}{RT^2}=\frac{nd\ln K}{dT}.$$

The temperature coefficient of an equilibrium constant expressed in terms of concentrations rather than partial pressures is given by

$$\frac{d\ln K}{dT}=\frac{\Delta U}{RT^2},$$

where ΔU is the change in internal energy in the reaction. (See, for example, reference [**5**], p. 147). Hence between the last two equations we obtain the following important relationship between the activation energies and the internal energy change:

$$E-E'=n\Delta U.$$

This may be compared with the following equation, which is often quoted, though it is not strictly correct:

$$E-E'=\Delta H.$$

As noted previously, the positive number n is often equal to unity and can always be made so by appropriate choice of the stoichiometric equation. On the other hand, ΔU and ΔH are *not* equal, except when the number of molecules is the same on each side of the chemical equation. For the perfect gas reaction

$$aA+bB=cC,$$

we have

$$\Delta U=\Delta H-RT(c-a-b).$$

The difference between ΔU and ΔH is often rather small, but they are not generally identical and this point needs to be borne in mind.

2.6 Parallel and consecutive reactions

Most reactions are very far from being simple. For many years the reaction

$$H_2+I_2\rightleftharpoons 2HI$$

was taught to students as the classic example of a simple bimolecular reaction involving stable species. It was then shown that even this reaction proceeds by a more complicated mechanism [**6**]. It should be noted that a given form of kinetics (in the case of the above reaction second-order kinetics) may be displayed by a variety of mechanisms. There is no unique connection between the *stoichiometry* of a reaction, its *kinetics*, and the reaction *mechanism*.

The great majority of reactions involve two or more elementary processes taking place in parallel or consecutively. In the latter case the overall reaction takes place through 'intermediates'; these may be stable substances which can be isolated, or they may be highly unstable and transient species such as free atoms or radicals.

A good example of parallel reactions is provided by the two modes of decomposition of an alcohol, e.g. ethyl alcohol:

$$C_2H_5OH = C_2H_4 + H_2O,$$

$$C_2H_5OH = CH_3CHO + H_2.$$

In general these occur simultaneously. The relative amount of ethylene and acetaldehyde obtained therefore depends on the relative speeds of the two reactions competing with each other for the available alcohol. These speeds in their turn are determined by the choice of catalyst and temperature.

Important instances of consecutive reactions occur in substitution processes, e.g.

$$CH_4 + Cl_2 = CH_3Cl + HCl,$$

$$CH_3Cl + Cl_2 = CH_2Cl_2 + HCl, \text{ etc.},$$

and also very frequently in oxidation processes where the desired product may oxidize further to give an undesired product. For example, in the oxidation of methanol the desired formaldehyde is readily oxidized further to carbon dioxide:

$$CH_3OH \rightarrow HCHO \rightarrow CO_2.$$

The formation of resins, tarry matter, etc., by consecutive reaction is also a common situation in organic chemistry.

The formal kinetics of parallel and consecutive reactions are necessarily more complicated than those of simple reactions. Consider for instance, the competing reactions:

$$A + B \rightarrow C, \qquad A + B \rightarrow D,$$

and let it be supposed that both reactions are first order with respect to A and B and that both are essentially irreversible. The rates of formation of C and D respectively will be given by

$$r_C = k_1[A][B], \tag{2.23}$$

$$r_D = k_2[A][B], \tag{2.24}$$

whilst the rate of consumption of A (or B) will be the *sum* of these expressions:

$$r_A = k_1[A][B] + k_2[A][B]. \tag{2.25}$$

Similar considerations apply to consecutive reactions. Let it be supposed that the pair of reactions

$$A+B \rightarrow C, \qquad A+C \rightarrow D,$$

are first order with respect to each of the substances involved and also that they are essentially irreversible. As in the example of the parallel reactions discussed above, the rate of consumption of A is the sum of the rates of the two reactions, but the rate of formation of C is their difference. Thus

$$r_B = k_1[A][B], \tag{2.26}$$

$$r_D = k_2[A][C], \tag{2.27}$$

$$r_A = k_1[A][B] + k_2[A][C], \tag{2.28}$$

$$r_C = k_1[A][B] - k_2[A][C], \tag{2.29}$$

where the subscripts denote consumption or formation of particular substances.

It will be seen that in a batch process the rate of formation of C is initially positive but, as reaction proceeds, it passes through zero and becomes negative (due to the second term on the r.h.s. of equation (2.29) becoming larger than the first) and finally approaches zero. The corresponding concentration of the substance C passes through a maximum and subsequently declines to zero. This substance is essentially a *reaction intermediate* as far as the formal kinetics are concerned, although it may be the substance which is desired as the useful product.

Example 2.3

A substance A decomposes to form a product P, which itself reacts to form a worthless material M. It is desired to produce P in a dilute solution by using a batch reactor. Both reactions are irreversible and first order. Determine the maximum fraction of A obtainable as P, and the reaction time at which this occurs.

Solution

We have

$$A \xrightarrow{k_1} P \xrightarrow{k_2} M$$

occurring in a constant-volume (since the solution is dilute) batch reactor. Application of equations (2.1) and (2.2) to this particular case gives us (see Example 2.1)

$$\frac{dc_A}{dt} = -k_1 c_A, \tag{2.30}$$

and

$$\frac{dc_P}{dt} = k_1 c_A - k_2 c_P. \tag{2.31}$$

Equation (2.30) is simply integrated, using the initial condition that $c_A = c_A^0$ at $t = 0$, to give

$$c_A = c_A^0\, e^{-k_1 t}. \tag{2.32}$$

This is then substituted into equation (2.31), which when integrated gives

$$c_P = A\, e^{-k_2 t} + \frac{k_1 c_A^0}{k_2 - k_1}\, e^{-k_1 t}. \tag{2.33}$$

Using the initial condition, $c_P = 0$ when $t = 0$, to determine the integration constant A we finally obtain

$$\frac{c_P}{c_A^0} = \frac{k_1}{k_2 - k_1}\left(e^{-k_1 t} - e^{-k_2 t}\right). \tag{2.34}$$

We can see from equation (2.34) that the fraction of A obtainable as P, c_P/c_A^0, passes through a maximum and decreases to zero as the reaction time becomes very large. Clearly the maximum will occur when $dc_P/dt = 0$. Differentiating equation (2.34) we obtain

$$\frac{dc_P}{dt} = 0 \quad \text{when} \quad k_1\, e^{-k_1 t} = k_2\, e^{-k_2 t}, \tag{2.35}$$

i.e

$$t_{\max} = \frac{\ln (k_2/k_1)}{k_2 - k_1} = 1/\text{log mean } k. \tag{2.36}$$

The value of c_P at $t_{\max}$ is obtained by substituting equation (2.36) in equation (2.34). The result is

$$\left(\frac{c_P}{c_A^0}\right)_{max} = \left(\frac{k_1}{k_2}\right)^{k_2/(k_2 - k_1)} \tag{2.37}$$

It can be shown from these answers that, for a given value of k_1, as k_2 decreases the maximum fraction of A obtainable as P increases and the time at which this occurs increases, too.

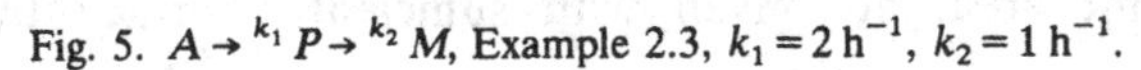

Fig. 5. $A \rightarrow^{k_1} P \rightarrow^{k_2} M$, Example 2.3, $k_1 = 2\ \text{h}^{-1}$, $k_2 = 1\ \text{h}^{-1}$.

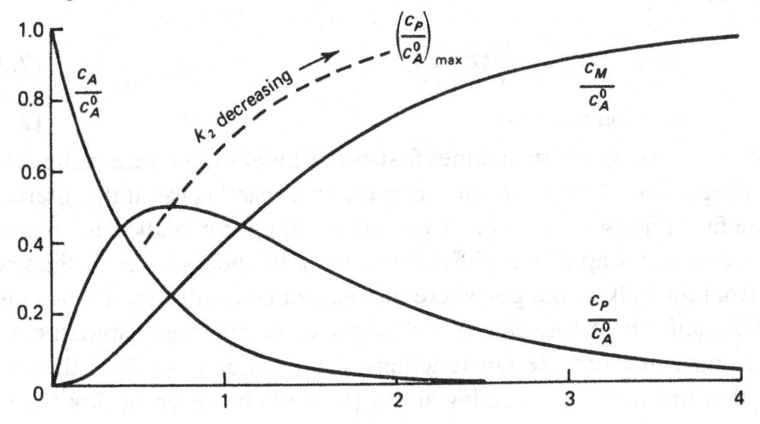

Fig. 5 shows c_A/c_A^0, c_P/c_A^0 and c_M/c_A^0 as functions of time in a constant-volume batch reactor. The curves have been drawn for $k_1 = 2\ h^{-1}$, $k_2 = 1\ h^{-1}$. Equation (2.36) gives $t_{max} = 0.69$ h and equation (2.37) shows that at this time $(c_P/c_A^0)_{max} = 0.50$. The dotted line shows how $(c_P/c_A^0)_{max}$ varies with t_{max} as k_2 is decreased. Note that the curve for (c_A/c_A^0) is unaffected by changing k_2.

We shall have more to say later about parallel and consecutive reactions. It commonly occurs that the designer has to consider how best to arrange matters so that of two competing (parallel) reactions the one giving the desired product is the faster. Again, if an intermediate component in a chain of consecutive reactions is the desired product, the designer must try to arrange reactor conditions to produce the maximum amount of the intermediate. This can often be achieved by an appropriate choice of the temperature or of the mixing conditions in the reactor.

2.7 Rate-limiting steps

When a process involves several consecutive steps – whether of reaction or diffusion, etc. – it is often said that one of them is rate-limiting. Yet paradoxically they may all be taking place at the same speed! It is essential for the student to have clear ideas about the meaning of the term 'rate-limiting', or 'rate-determining'.

The matter can be illustrated by reference to a process in which diffusion occurs, as well as chemical reaction. For example, if air were flowing through a heated carbon tube, a given molecule of oxygen would have to diffuse up to the carbon surface before it could react. Let it be supposed that the rate of diffusion through unit area, and the rate of reaction per unit area, are given with sufficient accuracy by the expressions:

$$\text{diffusion rate} = \frac{D}{x}(c - c_i), \tag{2.38}$$

$$\text{reaction rate} = kc_i. \tag{2.39}$$

The second expression assumes first-order kinetics, the rate being taken as proportional to the concentration c_i of the reacting gas at the interface. The first expression is familiar enough as an approximation in chemical engineering; it applies to diffusion across a boundary layer of thickness x from the bulk of the gas where the reagent concentration c is assumed to be uniform. We assume that this reactor is in a steady state, i.e. that the concentrations are not functions of time. Equation (2.1) then tells us that the molecules reacting at any place are balanced by flows to and from that place.

Since in this steady state as much gas reacts at the surface as diffuses up to it, the above rates are equal and this gives the means of eliminating the unknown surface concentration c_i. Thus equating the two rates we have

$$c_i = \frac{(D/x)c}{k+(D/x)}. \tag{2.40}$$

Hence, substituting in either of the two previous equations, we obtain:

$$\text{rate} = \frac{k(D/x)c}{k+(D/x)}. \tag{2.41}$$

It follows that if $D/x \gg k$, the expression reduces to

$$\text{rate} = kc \tag{2.42}$$

while if $D/x \ll k$ it reduces to

$$\text{rate} = \frac{D}{x}c. \tag{2.43}$$

In the former case (which tends to occur in a lower range of temperature) the observed rate is thus determined by the value of the velocity constant k and is said to be *chemically controlled.* In the latter case (which tends to occur in a higher range of temperature)* the factors which control the rate are D and x, and the process is said to be *diffusion controlled.*

Looking at the matter a little differently, from the standpoint of equation (2.40), the two limiting cases clearly correspond to c_i being (a) almost equal to c and (b) very small compared to c, respectively. Using language which is graphic rather than exact, one might say that in the former case the overall process waits on the chemical reaction at the surface and in the latter case it waits on the arrival of new molecules by diffusion.

It will be clear from this simple example that the question which of two or more successive steps is rate-controlling is a matter *not* of any difference in their actual rates (for these are equal at the steady state) but rather of a difference in the magnitudes of the *coefficients* of the process. If it were found by experiment that any factor which doubled k had the effect of doubling the observed rate, whilst factors influencing D/x had no effect, it could be concluded that the process is chemically controlled. And conversely.

In general, provided that the rate coefficients have the same dimensions (as with k and D/x in the above example), the rate-controlling step is the one whose coefficient is much the smallest. An equally simple state-

* In gases D varies as $T^{3/2}$ to T^2. Although D therefore increases with temperature, it is usually overtaken by the increase of k, since the latter is an exponential function of temperature.

ment would no longer be possible if the coefficients had unequal dimensions; this would occur, for example, if one process were first order in the reagent concentration and another were second order. Under these circumstances one process may be rate-controlling over a certain concentration range and a different process may be controlling over a different range.

But also, of course, even when the above condition concerning dimensionality is satisfied, it may occur that no single process is rate-controlling. Thus, in the above example, if k and D/x were approximately equal (i.e. in an intermediate range of temperature) neither of the limiting rate expressions, equation (2.42) or (2.43), would be applicable and equation (2.41) would have to be used instead.

The case of consecutive processes which has been discussed must be carefully distinguished from the case of *parallel routes.* Let it be supposed that a product D may be formed from a reagent A by two distinct paths*

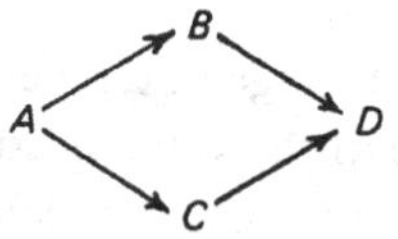

$$A \rightarrow B \rightarrow D, \quad A \rightarrow C \rightarrow D$$

In this situation it is clear that it is the *faster* of the two routes which has the greater influence on the overall rate of formation of D. In the limit, where the rate of the one route is very small compared to that of the other, its effect may be neglected. In short, when comparing *successive steps* we must pay attention to the one which is intrinsically the slowest and when comparing *parallel routes* we must attend to the one whose overall rate is the fastest.

Example 2.4

Consider the reaction system of Example 2.3. If now M is the desired product, derive the equation for c_M as a function of time. Hence show that if $k_1 \gg k_2$, the second reaction is rate-determining, whereas if $k_1 \ll k_2$, the first is rate-determining.

* For example

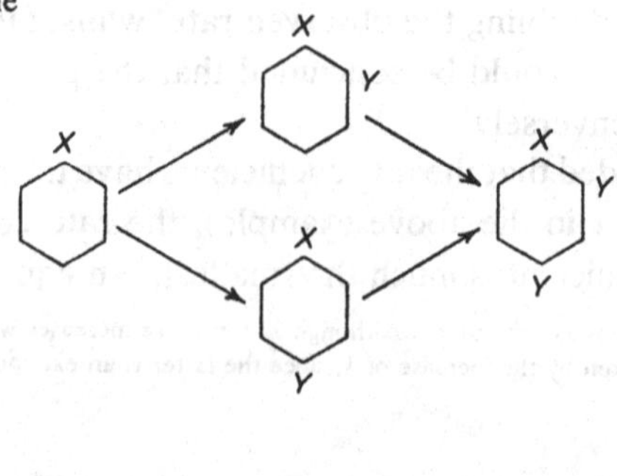

[Scheme: benzene ring → X-substituted ring → (via XY-ortho or XY-para isomers) → disubstituted product with X and Y]

Finally, it should be mentioned that the concept of a rate-determining step is not always applicable. Where a sequence of reactions, all essentially irreversible, occurs, then one may say that the sequence 'bottle-necks', at the rate-determining step. However, in more complicated reaction schemes involving, for example, steps which are reversible, and in which a given species may play more than one role in the scheme, no such convenient simplification as a 'rate-determining step' may be possible. If so, this may be unfortunate. The great value of being able to speak of a rate-determining step, in cases where this is valid, is that it readily focuses attention on those aspects of the process which should be examined with a view to improving the overall process rate.

The following examples may refresh the reader's memory of chemical kinetics, as well as illustrating some of the points of this chapter.

Example 2.5

In an experiment to determine the second-order reaction rate constant for the process

$A + 2B \rightarrow$ products,

an initial quantity of A is dissolved in a volume of solvent which does not fill a large stirred vessel. A flow into the vessel of B dissolved in the same solvent is started, and, at times measured from this point, the fraction of the original A remaining is determined. There is no discharge from the vessel, which may be considered uniformly mixed at all times.

From the table of measurements, demonstrate that a rate constant defined by

$-r_A = kc_Ac_B$

is consistent with these measurements, and evaluate the rate constant.

Initial concentration of A	2.73 kmol m^{-3}.
Concentration of B in added solution	5.47 kmol m^{-3}.
Initial volume of liquid in the vessel	12.0 m^3.
Volume flow rate of B solution	1.8 m^3 h^{-1}.

t (hours)	0	1.6	3.2	4.8	6.4	8.0	9.6	11.2	12.8	14.4
fraction of A remaining	1.0	0.91	0.76	0.61	0.47	0.35	0.25	0.17	0.11	0.07

[*Answer.* $k = 0.146$ m^3 $(kmol\ h)^{-1}$.]

Example 2.6

Acetylcholine (A) is hydrolysed to a product B by the enzyme esterase (E) through the formation of a complex (AE):

$A + E \rightleftharpoons AE \rightarrow B + E.$

Assuming that a stationary-state analysis is possible, show that the rate equation takes the Michaelis–Menten form

$-da/dt = ka/(K+a)$,

where k and K are constants and a is the acetylcholine concentration.

A solution containing acetylcholine is hydrolysed and its concentration followed as a function of time

Acetylcholine concentration (mol m^{-3})	t(s)
10.0	0.0
1.0	7.45×10^{-3}
0.5	8.57×10^{-3}
0.25	9.56×10^{-3}

Use these results to find K and k.

[*Answers.* $K = 2.39$ mol m^{-3}, $k = 1.92$ kmol m^{-3} s^{-1}.]

Example 2.7

A substance A in solution dimerizes to form a product B, which decomposes to give a product C. The kinetics were investigated in a well-stirred batch reactor. The concentrations of A, B and C were measured at given times: the results are shown below. What kinetics and velocity constants can you deduce from these results?

Compare the values of the concentration of C calculated from your kinetic parameters with the values shown in the table. Discuss the precision of your results.

Time (hours)		0	0.03	0.04	0.06	0.1	0.15	0.2	0.3
Concentrations (kmol m^{-3})	A	1	0.770	0.724	0.630	0.500	0.400	0.330	0.260
	B	0	0.107	0.121	0.138	0.130	0.105	0.085	0.050
	C	0			0.04		0.20		0.33

[*Answers.* $A \rightarrow B$ second order, $k = 4.95$ m^3 (kmol h)$^{-1}$, $B \rightarrow C$ first order, $k = 12.4$ h^{-1}.]

Example 2.8

The rate of conversion of trypsinogen into trypsin is found to follow a rate equation of the form

$$-\frac{dg}{dt} = kgs,$$

where g is the concentration of trypsinogen and s the concentration of trypsin.

In separate batch experiments at different temperatures the reaction rate was observed to pass through a maximum at a time t_{max} after the start of the experiment.

Temperature (K)	293	310
Initial concentration of trypsinogen (kmol m^{-3})	0.65	0.62
Initial concentration of trypsin (kmol m^{-3})	0.05	0.08
t_{max} (h)	2.40	0.434

Use these data to determine the activation energy of the reaction.

[*Answer.* 66 000 kJ $kmol^{-1}$.]

Example 2.9

A reagent A can be isomerized isothermally to a product B by a first-order irreversible reaction in the gas phase. A by-product, C, is also formed by the irreversible, second-order dimerization of A, which reaction also occurs isothermally in the gas phase.

The reaction was carried out in a laboratory constant-volume batch reactor at two different temperatures, using the same initial concentration of pure A, a_0, in each case. The reactions were allowed to go to completion, and the concentration of C, c_∞, was then found to be $0.117\, a_0$ at T_1, $0.0537\, a_0$ at T_2. Calculate the ratio $k_2 a_0/k_1$ at the two temperatures, where k_2 and k_1 are the second- and first-order velocity constants, based on the rate of destruction of A.

[*Answers.* 0.667 at T_1, 0.250 at T_2.]

Example 2.10

The irreversible liquid–phase reaction $A+B \rightarrow D$ is carried out adiabatically in a batch reactor with equal initial concentrations of A and B. Derive an integral expression for the variation of the concentration of A with time in terms of the initial temperature T_0, the corresponding rate constant k_0, the heat of reaction ΔH and the energy of activation E.

The initial concentrations of A and B are both 8 kmol m^{-3} and $T_0 = 90$ °C. Assuming the density and heat capacity of the system are those of water and remain essentially constant, what is the limiting temperature for the system? Estimate the time taken for the initial concentration of A to fall by (i) 25% and (ii) 50%, and calculate the corresponding reactor temperatures.

Data at 90 °C:

ΔH −88 500 kJ per kmol of A

E 77 500 kJ $kmol^{-1}$

k_0 7.39×10^{-5} m^3 $kmol^{-1}$ s^{-1}

[*Answers.* 259 °C, (i) 156 s, 132 °C, (ii) 181 s, 175 °C.]

Symbols

a_i	Activity of species i.
a, b, c	Stoichiometric coefficients in equations (2.14), (2.16) and (2.17).
c_i	Concentration of species i, kmol m^{-3}.
D	Diffusion coefficient, $m^2 s^{-1}$.
E	Activation energy, J $kmol^{-1}$.
H	Enthalpy, J $kmol^{-1}$.
k	Velocity constant, units depend on order of reaction.
K	Equilibrium constant.
n	Stoichiometric number in equations (2.16) and (2.17).
n_i	Number of moles of species i.
r	Reaction rate, kmol $m^{-3} s^{-1}$.
R	Gas constant, J $kmol^{-1} K^{-1}$.
t	Time.
T	Absolute temperature.
U	Internal energy, J $kmol^{-1}$.
V	Volume of reactor.
x	Boundary-layer thickness in equation (2.38).
α, β, γ	Reaction orders.
γ_i	Activity coefficient of species i.
ν_i	Stoichiometric number of species i.
ϕ	Factor in equation (2.19).
ξ	Extent of reaction in equations (2.9) and (2.10).

References

1. Kistiakowsky, G. B., *J. Amer. Chem. Soc.*, 1928, **50**, 2315.
2. Denbigh, K. G. and Prince, A. J., *J. Chem. Soc.*, 1947, 790.
3. Eckert, G. A. and Boudart, M., *Chem. Engng Sci.*, 1963, **18**, 144.
4. Mason, D. M., *Chem. Engng Sci.*, 1965, **20**, 1143.
5. Denbigh, K. G., *The Principles of Chemical Equilibrium* (4th edn, Cambridge University Press, 1981).
6. Sullivan, J. H. *J. Chem. Phys.*, 1966, **46**, 73.

   Several of the textbooks mentioned at the end of Chapter 1 have sections on chemical kinetics, as do general texts on physical chemistry, e.g.:

7. Atkins, P. W., *Physical Chemistry* (Oxford, 1978).

   There are many books on reaction kinetics, and three of different types are the following:

8. Pilling, M. J., *Reaction Kinetics* (Oxford, 1975).
9. Benson, S. W., *Thermochemical Kinetics: Methods for the Estimation of Thermochemical Data and Rate Parameters* (Wiley, 1976).
10. Bamford, C. H. and Tipper, C. F. H., (eds.), *Comprehensive Chemical Kinetics* (A many-volumed series by Elsevier.)

3

Tubular reactors

3.1 The plug flow assumption (P.F.A.)

As described already, the tubular reactor is a type of continuous flow reactor in which no attempt is made to induce mixing of fluid between different points along the overall direction of flow. It follows that the most appropriate first approximation for the purpose of calculating the reactor's performance is based on the assumption that there is plug flow (also called piston flow).

What is this assumption? Different authors give it slightly different meanings but plug flow will be defined here as being an idealized state of flow such that: (1) over any cross-section normal to the fluid motion the mass flow rate and the fluid properties (pressure, temperature and composition) are uniform; (2) there is negligible diffusion relative to the bulk flow. Reactors approximately satisfying this assumption will be called plug-flow reactors (P.F.R.s).

The effect of the restrictions under (1) is that all elements of fluid (which may be imagined as being contained within small envelopes) spend an equal time in passing through the reactor and pass through the same sequence of pressure, temperature and concentration changes. The effect of the restriction under (2) is that molecules of reagents and products do not diffuse from one such element of fluid to another during the passage of these elements through the reactor. It follows that reaction occurs to the same extent in each element and these can therefore be regarded as miniature batch reaction systems travelling through the system. The performance of the tubular reactor as a whole will thus be exactly the same as for a batch system having a duration equal to the time of passage through the tubular reactor and the same temperature and pressure history. However, it is in connection with the pressure history that some care has to be taken, for (as was noted earlier in §2.3) it must not be assumed that the volume of an element of fluid necessarily remains constant. The method of calculation now to be described takes care of this effect.

3.2 Elementary design method

The reactor will be assumed for the moment to contain no solid catalyst or any sort of packing. In Fig. 6 let P and Q be two planes containing between them an infinitesimal part dV_r of the total reactor volume V_r. If we apply the mass balance equation (equation (2.1)) to this element, we have

$$\text{Moles in} = \text{Moles out} + \text{Moles Reacted} + \text{Change of moles in element,}$$

in which the last term will be zero if we assume that the reactor is in a steady state. In accordance with the plug-flow assumption all concentrations are uniform over the cross-section, as are the flow rate and reaction rate. Finally let r be the reaction rate expressed as moles of product made per unit time and volume (we *could* choose the rate of destruction of reagent). Thus we can write

$$d\,(\text{molar flow of product}) = r\,dV_r. \tag{3.1}$$

The molar flow at any cross-section of the P.F.R. can be expressed in a variety of ways. One obvious choice is the volumetric flow rate, v, times the concentration, c; another is the mass flow rate, G, times the variable y, the moles of product per unit mass flow. A third choice might be the molar rate of *feed*, m_f, times the fractional conversion, x. Thus we can write equation (3.1) as

$$d(vc) = d(Gy) = d(m_f x) = r\,dV_r. \tag{3.2}$$

For a liquid-phase reaction, it is almost always acceptable to assume that v is a constant, but for a gas-phase reaction it frequently is not. Although there is a pressure drop across the reactor (to produce the flow through it), such a pressure drop is usually small in comparison with

Fig. 6. Volume element of a tubular reactor.

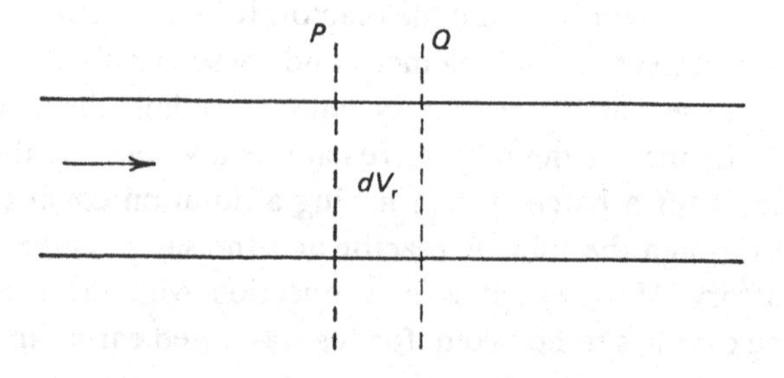

the mean pressure, particularly for high-pressure reactors. Thus it is usual to model a P.F.R. as a constant-pressure reactor, and for a gas-phase reaction which results in a change in the number of moles, the volume flow rate will change.

It would thus seem simplest to choose the mass flow rate G, in which case $d(Gy) = G\,dy$, and equation (3.2) can be integrated to give

$$V_r = G \int_{y_i}^{y_e} \frac{dy}{r}. \tag{3.3}$$

The inlet value of y, y_i, is usually zero, but if, for example, there is a recycle stream, there may be some product in the reactor feed, and y_i will then be non-zero.

Similar equations can be readily derived for application to reactors containing particles of solid catalyst. Let r' be the reaction rate per unit mass of catalyst and let W_r be the total mass of catalyst necessary to bring the exit concentration up to a value y_e. Then analogously to equation (3.3) above we have

$$W_r = G \int_{y_i}^{y_e} \frac{dy}{r'}. \tag{3.4}$$

The usefulness of the above equations is necessarily limited by the applicability of the P.F.A. Under certain circumstances this assumption can be used quite reliably; under others, as will be shown shortly, it leads to gross errors. Nevertheless, these equations are an invaluable starting point for the discussion of tubular reactor performance.

In order to carry out the integration in equations (3.3) or (3.4) it is necessary to know the reaction rate r, or r', in relation to the variable y. Consider, for example a homogeneous gaseous reaction between reagents A and B, the reaction being supposed to be virtually irreversible and to be of the α order with respect to A and of the β order with respect to B.

Then, as in equation (2.7) we have

$$r = k[A]^\alpha [B]^\beta, \tag{3.5}$$

and inserting this in equation (3.3), assuming no product in the feed stream,

$$V_r = G \int_0^{y_e} \frac{dy}{k[A]^\alpha [B]^\beta}. \tag{3.6}$$

We have now to relate the volume concentrations $[A]$ and $[B]$ to the variable y, and it is here that some of the simplicity of basing the flow on the constant quantity G (instead of v, which in general is not constant) is lost. We shall show how to do this later in examples.

The only remaining problem before integration can be carried out concerns the velocity constant k. This depends very sensitively on the temperature but there are two limiting cases which allow equation (3.6) (or the corresponding equation for a packed reactor as obtained from equation (3.4)) to be evaluated fairly simply. These are:

(*a*) The case where the temperature may be assumed to remain constant along the length of the reactor (as well as over the cross-section as is required by the P.F.A.). In this instance the velocity constant can be taken outside the integral sign, and the integration can be carried out forthwith – either analytically in suitable instances or otherwise numerically or graphically, as will be described in Appendix I to this chapter, which concerns an example of the oxidation of NO to NO_2. The condition most obviously favouring the assumed constancy of temperature is a negligibly small heat of reaction. Otherwise approximate constancy may still be obtained if the reactor wall is held at constant temperature (e.g. by means of a jacket) and if the reactor diameter is small enough, or if the turbulence of the reacting fluid is great enough, to ensure that the heat of reaction is transferred very effectively from within the body of the fluid up to the wall.

(*b*) The case where reaction takes place adiabatically, the walls of the reactor being so efficiently insulated that there is negligible heat loss in directions at right-angles to the direction of flow. In this instance the temperature will rise or fall along the length of the reactor, according to whether the reaction is exothermic or endothermic respectively, and in a manner which is readily calculated from a knowledge of the heat of reaction. For this purpose an energy balance is set up relating the temperature change between the reactor inlet and a given cross-section to the value of the variable y, which measures the extent of reaction at the same cross-section. It is supposed that conduction of heat in the direction of flow can be neglected. The velocity constant (which is assumed to be known as a function of temperature) thereby becomes known as a function of y, and the required integration of the preceding equations can be carried through, either numerically or graphically. This will be described in Appendix II to this chapter where the oxidation of NO to NO_2 is again used as an example, but in this case an adiabatic reactor is assumed.

3.3 Variable volume flow rate

In the previous section the proper application of the mass balance equation to a tubular reactor was shown to lead to equation (3.3). In systems involving no change of density of the fluid stream as it passes through the reactor, the time required for the fluid stream to reach any cross-section is equal to the void volume up to that cross-section divided by the volumetric feed rate. It then becomes possible to adapt equation (2.3), which refers to a *constant-volume batch* reactor, by substituting $v\,dc/dV_r$ for dc/dt. Although this is correct in this instance it is very misleading. To substitute for 'time' in the batch reactor equation to obtain a continuous reactor expression is not correct in general and can lead to error and confusion. For example, in a continuous reactor in a non-steady state, time enters 'of its own right', being involved in term 4 of the mass balance equation, equation (2.1).

When there *is* a change in density of the fluid stream as it passes through the reactor, then the substitution of $v\,dc/dV_r$ for dc/dt is not only *misleading*, it is *incorrect.*

As an example, let us take an isothermal nth order irreversible gas-phase reaction of a single initially-pure reagent, in which each mole of reagent produces $(1+\varepsilon)$ moles of product. The molar flow rate of reagent at any point in the reactor where the conversion is x is given by $m_f(1-x)$. The *concentration* of the reagent, c, at that point is given by the mole fraction times the total molar concentration. Hence

$$c=\frac{(1-x)}{(1+\varepsilon x)}\cdot\frac{P}{RT}, \tag{3.7}$$

provided we can assume ideal gas behaviour. Here P is the pressure, R the gas constant, and T the temperature. Substituting equation (3.7) in the nth order expression for r, and integrating equation (3.2), we obtain

$$V_r=\frac{m_f}{k}\left(\frac{RT}{P}\right)^n\int_0^{x_{\text{out}}}\left(\frac{1+\varepsilon x}{1-x}\right)^n dx, \tag{3.8}$$

which relates the conversion to product, x_{out}, of the stream leaving a reactor to the volume V_r.

To illustrate the effect of ε, Fig. 7 shows dimensionless volume as a function of x for $n=2$, a second-order reaction. The examples chosen are

(*a*) $\varepsilon=-\frac{1}{2}$, e.g. a dimerization,
(*b*) $\varepsilon=0$, e.g. an isomerization,
(*c*) $\varepsilon=1$, e.g. a decomposition into two molecules.

For a conversion of 90%, the dimensionless volume for $\varepsilon=0$ is 9. This is reduced to 3.63 if $\varepsilon=-\frac{1}{2}$, and increased to 27.7 if $\varepsilon=1$. Clearly for P.F.R. design one cannot in general use procedures appropriate for constant-volume batch reaction. Such procedures, for $\varepsilon=0$, are incorrect for two reasons. Firstly the gas (at approximately constant pressure, be it remembered) speeds up or slows down as it passes through the reactor.

Thus a simple relation between volume and time in the reactor is ruled out. As well as this, the kinetics are changed by such contraction or expansion, so the constant-volume reactor is a kinetically inappropriate analogue as well.

Fig. 7. Conversion versus P.F.R. volume. Second-order reaction with change in number of molecules.

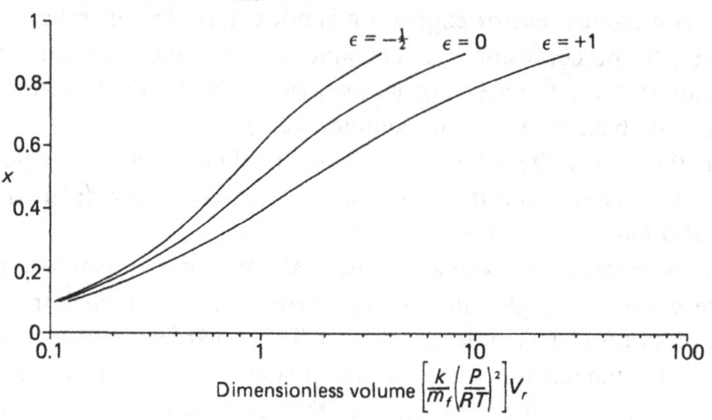

Example 3.1*

Under appropriate conditions, acetaldehyde vapour reacts to give methane and carbon monoxide by the reaction

$$CH_3CHO \xrightarrow{k} CH_4 + CO.$$

0.1 kg s^{-1} of acetaldehyde vapour is to be decomposed at 520 °C and 1 atm in a tubular reactor. The reaction under these conditions is known to be irreversible and second-order with respect to CH_3CHO. The velocity constant, k, is 0.43 m^3 $(kmol)^{-1}$ s^{-1}. What will be the volume of reactor required (*a*) for 35% decomposition of the feed acetaldehyde, and (*b*) for 90% decomposition?

Solution

In the terminology of equation (3.8) we have, in this case, $n = 2$ and $\varepsilon = 1$, and thus equation (3.8) becomes

$$V_r = \frac{m_f}{k}\left(\frac{RT}{P}\right)^2 \int_0^{x_{out}} \left(\frac{1+x}{1-x}\right)^2 dx. \tag{3.9}$$

The integral in equation (3.9) can be evaluated analytically, the result being

$$\frac{4}{1-x_{out}} - 4 + 4\ln(1-x_{out}) + x_{out},$$

* This problem is derived from a Cambridge University Chemical Engineering Tripos question.

which has the value 0.78 for 35% conversion, and 27.7 for 90% conversion. At 520 °C and 1 atm, $(RT/P) = 65.1\ \text{m}^3\ (\text{kmol})^{-1}$. Also $m_f = 0.1\ \text{kg s}^{-1} \equiv 0.1/44$ kmol acetaldehyde feed per second. Substituting in equation (3.9), we obtain

$$V_r = \frac{0.1}{44 \times 0.43} \times (65.1)^2 \times 0.78 = 17.5\ \text{m}^3$$

for 35% conversion. Similarly the volume required for 90% decomposition is $619\ \text{m}^3$.

***Example* 3.2**

For the above example, work out the times required for a constant-volume batch reactor at 520 °C to decompose 35% and 90% of a pure acetaldehyde vapour feed initially at 1 atm. Using the (incorrect) substitution of $t = V_r/v$, calculate the equivalent continuous reactor volumes for the above feed rate of $0.1\ \text{kg s}^{-1}$. Compare with the correct answers.

[*Answers.* 81.4 s, 12.0 m^3; 1360 s, 201 m^3.]

3.4 Residence-time, space-time, space-velocity

These terms are frequently used in discussing the performance of a reactor, and can be conveniently considered here. The *residence-time* of an element of fluid leaving a reactor is the length of time spent by that element within the reactor. For a tubular reactor, under plug-flow conditions, the residence-time is the same for all elements of the effluent fluid. We shall consider situations where this is not the case in a later chapter, but here we shall be concerned only with plug-flow conditions.

The *space-time* is defined as the 'reactor volume' divided by the 'volume rate of flow' of the fluid, and thus has the dimensions of time. If by the 'reactor volume' one means the void volume, and by the 'volume rate of flow' one means that measured at the temperature and pressure of the reactor, then the space-time is equal to the residence-time for an isothermal tubular reactor in which there is no change of density of the process fluid. The *space-velocity* is simply the reciprocal of the space-time.

There is thus a simple physical interpretation of the space-velocity and the space-time in this particular case. A space-velocity of $10\ \text{h}^{-1}$, for example, would mean that the fluid in the reactor was 'replaced' ten times per hour, and each element would, on average, stay in the reactor for six minutes (a space-time of six minutes).

In cases where the fluid density does change in the reactor, things are not so simple. The 'volume rate of flow' must be referred to some standard condition, which need be at neither the temperature nor the pressure of any part of the reactor. Again, the 'reactor volume' may be the *total* reactor volume, including that of a catalyst packing. Space-time and space-velocity then cease to have any simple physical meaning, but they retain a utility in characterizing reactor performance, as will be seen later.

Example 3.3

For the acetaldehyde decomposition example of §3.3 calculate the *residence-time* in the reactor for 35% and for 90% decomposition. Why are these not the same as the answers to Example 3.2?

[*Answers.* 98 s and 2370 s.]

Example 3.4

Calculate the *space-time* and *space-velocity* for the conditions of Example 3.1, using the volume flow rate of the feed at the temperature and pressure of the reactor as the standard condition.

Solution

In Example 3.1 we saw that the two reactor volumes required were 17.5 m^3 for 35% decomposition, and 619 m^3 for 90% decomposition. The volume rate of flow of the feed at the temperature and pressure of the reactor is given by $v = m_f(RT/P)\ m^3\ s^{-1}$, where m_f = kmol acetaldehyde feed s^{-1}. Substituting for m_f and RT/P we have

$$v = \frac{0.1}{44} \times 65.1 = 0.148\ m^3\ s^{-1}.$$

The space-time is

$$\frac{17.5}{0.148} = 118\ s \quad \text{for 35\% decomposition,}$$

and

$$\frac{619}{0.148} = 4190\ s \quad \text{for 90\% decomposition.}$$

The *space-velocity* is the inverse of the appropriate space-time.

We thus have three quantities of the dimensions of time and it is necessary to be clear why they are different in this case, as they will be in general. For 35% decomposition of acetaldehyde the figures are as follows:

81.4 s for a constant-volume batch reactor (Example 3.2),
98.0 s for the residence-time in a tubular constant-pressure reactor (Example 3.3),
118 s for the space-time in a tubular reactor, based on inlet conditions (Example 3.4).

The reason why these times are different (and they are much more different for 90% decomposition) is that this reaction involves a decrease in the density of the fluid stream, which will therefore accelerate as it passes through the reactor. The constant-volume batch reactor does not allow this dilution of the reagents by expansion. It therefore gives the shortest time for a given extent of reaction.

The space-time calculation gives the time that the process stream would spend in the reactor if it did not accelerate, while the residence-time takes the acceleration into account and is therefore the smaller of the two. Both use the actual reactor volume and this has to take the expansion, and hence dilution, into account. Therefore both are larger than the batch reactor figure.

If the reaction did *not* involve a change in density, then all three times would be the same. If this reaction were first order, the difference between the three times would still occur but would be less marked.

It may be asked why space-time and space-velocity are used, when they may have no simple physical interpretation. There are two valuable uses for these concepts. First, if the reactor conditions are varied, either within a set of experiments or within a set of calculations, it is convenient to correlate output in terms of the reactor volume divided by the volume feed rate measured under some standard conditions (which then of course cannot be the varying reactor conditions). The mass flow rate would suffice, but the volume flow rate may enable an order of magnitude picture of the 'time in the reactor' to be retained. Sometimes this picture is greatly distorted: the volume of *liquid* feed to a plant in which this liquid is vaporized and reacted has been used to define the space-time in the reactor. The reader will appreciate why this gave a space-time vastly greater than the residence-time.

The second use is where the kinetics of the reaction are unknown and where the space-time may be the only time which is calculable. This will occur in, for example, process investigational work on new reactions. In such work it is usually desirable to try to match in experimental work conditions which may obtain on the projected plant. This is particularly so if a bed of solid catalyst is to be used. Thus experiments of the classical chemical type (constant-volume batch reactions) may not be acceptable to establish a useful mathematical model of the overall kinetics on the plant.

Let us revert to the acetaldehyde decomposition example worked so far. It will be seen that we have used the known kinetics to work out the three 'times' mentioned. If the kinetics had been unknown this could not have been done. Supposing we had a tubular reactor of volume V_r, how could the kinetics of the reaction be determined?

The procedure would be to carry out experiments with this reactor at varying feed rates, measuring the extent of reaction of the stream leaving the reactor. Since the density of the fluid stream changes with the extent of reaction, we cannot say with any precision what the residence-time of the fluid in the reactor is, since its speed through the reactor increases from entry to exit. We do, of course, know the space-time, since that is based on the known reactor volume and known feed rate to the reactor.

One possible method might be to add 'inert' gas to the acetaldehyde vapour in such quantity that the change in density between entry and exit of the reactor could be neglected. In that case the batch reactor time and the residence-time would both be equal to the space-time. However, it is generally unwise in kinetic studies, especially involving solid catalysts, to change the nature of the process stream so much, if results valuable for practical use are to be obtained (i.e. in the absence of inert gas).

Using the results of such experiments, we can apply equation (3.8) to determine n and k (ε will be known from the stoichiometry). Our results will be in the form of values of x_{out} for various values of the feed rate, m_f. Defining the space-time, τ, as the reactor volume divided by the volumetric feed rate, we have, from equation (3.8),

$$\tau = \frac{V_r}{m_f}\left(\frac{P}{RT}\right) = \frac{1}{k}\left(\frac{RT}{P}\right)^{n-1} \int_0^{x_{out}} \left(\frac{1+\varepsilon x}{1-x}\right)^n dx. \qquad (3.10)$$

From our experiments we should be able to draw a curve of τ against x_{out}, the slope of which, according to equation (3.10) should be

$$\frac{d\tau}{dx_{out}} = \frac{1}{k}\left(\frac{RT}{P}\right)^{n-1} \left(\frac{1+\varepsilon x_{out}}{1-x_{out}}\right)^n. \qquad (3.11)$$

Taking the logarithm of both sides of equation (3.11), we obtain

$$\ln\left(\frac{d\tau}{dx_{out}}\right) = \ln\left[\frac{1}{k}\cdot\left(\frac{RT}{P}\right)^{n-1}\right] + n \ln\left(\frac{1+\varepsilon x_{out}}{1-x_{out}}\right); \qquad (3.12)$$

so k and n can be obtained from the intercept and slope of the appropriate log–log plot.

Of course this approach requires that the experiments be isothermal (k and T outside the integral in equation (3.10)). The values of k and n are likely to be much affected by experimental error and by the difficulty of measuring the slopes accurately. It would probably be best to use equation (3.12) to decide the choice of n (which may often be presumed to be an integer). Given the value of n we can use the experimental results directly with equation (3.10) to determine k.

If the reactor is not isothermal, then (3.10) must be written as

$$\tau = \frac{V_r}{m_f}\left(\frac{P}{RT_{\text{in}}}\right) = \frac{1}{T_{\text{in}}}\left(\frac{R}{P}\right)^{n-1} \int_0^{x_{\text{out}}} \frac{T^n}{k}\left(\frac{1+\varepsilon x}{1-x}\right)^n dx, \qquad (3.13)$$

where T_{in} is the temperature of the feed into the reactor.

Now to obtain τ (or V_r) for a desired x_{out}, we must know how T, and k, vary with the conversion x. If the reactor is adiabatic (as mentioned in §3.2 above) we can use the thermal balance to provide the connection between x and T.

The reader is advised to work through the two appendices at this stage.

The following three examples, of which the last is modelled on a problem due to Hougen, Watson and Ragatz [**1**], are straightforward applications of the principles involved in this chapter so far.

Example 3.5

N_2O_5 vapour undergoes the reaction

$$2N_2O_5 = 4NO_2 + O_2,$$

and this is first order and essentially irreversible. The N_2O_5 vapour passes through a tubular reaction vessel in which the pressure and temperature are almost constant. The flow rate at the inlet is n_0 moles per unit time.

If there is approximately plug flow in the vessel, show that the fraction F of the N_2O_5 which decomposes during passage is given by

$$\frac{kV_r}{V_0 n_0} = 2.5 \ln \frac{1}{(1-F)} - 1.5\,F,$$

where $r = k[N_2O_5]$ is the moles of N_2O_5 destroyed per unit time and volume, V_r is the volume of the reaction vessel and V_0 is the molar volume at the prevailing temperature and pressure. (Neglect the presence of N_2O_4.)

Example 3.6

The gas reaction $A = \nu B$ takes place catalytically, and its rate (kmols of A destroyed per unit time per unit overall volume of reactor) is proportional to the surface area of the catalyst per unit overall volume and to the concentration of the gas A.

The process takes place in a tubular reactor, packed with the catalyst, at approximately constant temperature and pressure. Assuming plug flow, show that the fraction of A which reacts during passage is given by

$$\frac{kSV_r}{V_0 n_0} = \nu \ln \frac{1}{(1-F)} - (\nu - 1)F,$$

where S is the surface area of catalyst per unit overall volume and k is a velocity constant per unit area. V_0 and n_0 are as in Example 3.5 and V_r is the total volume of the reactor.

***Example* 3.7**

The decomposition of phosphine takes place according to the stoichiometric equation

$4PH_3 = P_4 + 6H_2$.

The reaction is not appreciably reversible and is endothermic. It is first order, the rate being proportional to the phosphine concentration, and the temperature dependence of the velocity constant is given by the following equation (from International Critical Tables):

$$\log_{10} k = -\frac{18963}{T} + 2\log_{10} T + 12.130,$$

where $k[PH_3]$ is the moles of PH_3 destroyed per unit time and volume, k is in s^{-1} and T is the temperature in K.

It is proposed to produce phosphorus by the decomposition of phosphine at a feed rate of $16\ kg\ h^{-1}$ in a tubular reactor operating at atmospheric pressure. The highest temperature which can be used in the available material of construction is 680 °C and at this temperature the phosphorus produced is a vapour.

Calculate (a) the fractional conversion which might be expected in a tubular reactor of volume $1\ m^3$ if the temperature is held constant at 680 °C; (b) the fractional conversion in the same reactor operating adiabatically with an inlet temperature of 680 °C.

The heat of reaction at 18 °C is $\Delta H = 23\,720\ kJ\ kmol^{-1}$ of phosphine. Heat capacities in the required temperature range are as follows:

$P_4(g)$ $62.3\ kJ\ °C^{-1}\ kmol^{-1}$,

$PH_3(g)$ $52.6\ kJ\ °C^{-1}\ kmol^{-1}$,

$H_2(g)$ $30.1\ kJ\ °C^{-1}\ kmol^{-1}$.

[*Answers*, (a) 0.68; (b) 0.13.]

Further examples are given after the appendices at the end of this chapter.

3.5 Deviations from the plug-flow assumption

There are three distinct types of deviation from the uniformity over a cross-section which is postulated by the P.F.A.:

(a) The existence of temperature gradients at right-angles to the direction of flow, due to the heat of reaction.

(b) Diffusion from one fluid element to another, diffusion here being understood as including the effects due to turbulence or to the influence of the packing, as well as ordinary molecular diffusion and the convective mixing due to temperature differences.

(c) The existence of velocity gradients normal to the direction of flow.

In the case of reactors packed with a fixed bed of catalyst, the first of these factors is usually by far the most important and it will be discussed in more detail in the next section. Diffusion is also an important factor. The diffusion at right-angles to the direction of flow ('radial diffusion') must certainly be allowed for whenever an attempt is made to allow for the temperature gradients in the same direction. On the other hand, diffusion in the same direction as the flow ('longitudinal diffusion') is usually a good deal less significant. As regards the influence of velocity gradients, this is not usually a very important factor, except in the special case of unpacked tubular reactors when they operate under conditions of laminar rather than of turbulent flow. The reader will appreciate, however, that these statements are rough generalizations and they may not be applicable to tubular reactors of unusual design. An exceptionally small length/diameter ratio, for example, would tend to increase the influences both of longitudinal diffusion and of velocity gradient. A fuller discussion of factors (*b*) and (*c*) above will be given later (Chapter 5).

3.6 Transverse temperature gradients: a general discussion

The occurrence of temperature gradients in the *same* direction as the flow is by no means contrary to the conditions postulated by the P.F.A., and in Appendix II to this chapter an example is quoted showing how such gradients can be allowed for within the scope of this assumption. On the other hand, the existence of gradients at right-angles to the direction of flow (henceforth to be called radial or, better, transverse gradients) tends to invalidate the P.F.A. and whenever these gradients are appreciable the elementary design method described in § 3.2 above is of very little value.

The circumstances affecting the character of the transverse gradients may be briefly described. If an exothermic reaction takes place in a tube-shaped reactor from which heat is being removed through the wall by means of external cooling, the temperature profile may obviously be expected to be somewhat as is shown in Fig. 8(*a*), i.e. the reacting fluid will be hotter near the tube centre than it will be near the wall. At first sight, therefore, it might appear that transverse gradients could not occur if there were no transverse removal of heat, i.e. in a reactor having walls which were perfectly insulated. However, the effect of the velocity gradient needs also to be considered. In so far as the fluid near the tube centre travels faster than it does near the wall, its temperature rise, at a given distance along the tube, is less (since less reaction has occurred) and thus the shape of the temperature profile tends to be reversed (Fig. 8(*b*)). (This may be understood most clearly by visualizing a cold stream of reactant gas entering the reactor and giving rise to a cold plume driving through its centre.) Therefore when the effects of wall heat transfer and of velocity gradient operate simultaneously they might, under rather special circumstances, give rise to a more complex kind of temperature profile such as is indicated schematically in Fig. 8(*c*). However, the most commonly observed profiles obtained with exothermic reactions in externally cooled reactors are undoubtedly those of the sort illustrated in Fig. 8(*a*). Experimental results of this kind were presented earlier, in Fig. 3.

The reason why the elementary design method is erroneous when the transverse gradients are appreciable arises from the extreme sensitivity of reaction rate to changes of temperature. Consider, for example, the implications of Smith's temperature measurements as given in Fig. 3. These were obtained in a tubular reactor in which sulphur dioxide oxidation was taking place. With the tube wall maintained at 197 °C, it was found that the temperatures near the tube axis were as much as 300 °C higher. The activation energy for this reaction being about 8×10^7 J kmol^{-1}, it follows that the values of the velocity constant at the 'hot spot' and near the wall are approximately in the ratio

$$\frac{e^{-8\times10^7/770R}}{e^{-8\times10^7/470R}} \simeq 3000.$$

This simple calculation has been presented in order to emphasize what is undoubtedly the biggest factor of uncertainty in the design of fixed-bed reactors at the present time. This uncertainty arises from the difficulty of making a sufficiently exact calculation of the transverse temperature gradients under any given circumstances. The radial temperature variation found in Smith's experiments is probably rather exceptionally large,

Fig. 8. Types of temperature profiles obtained in exothermic reaction.

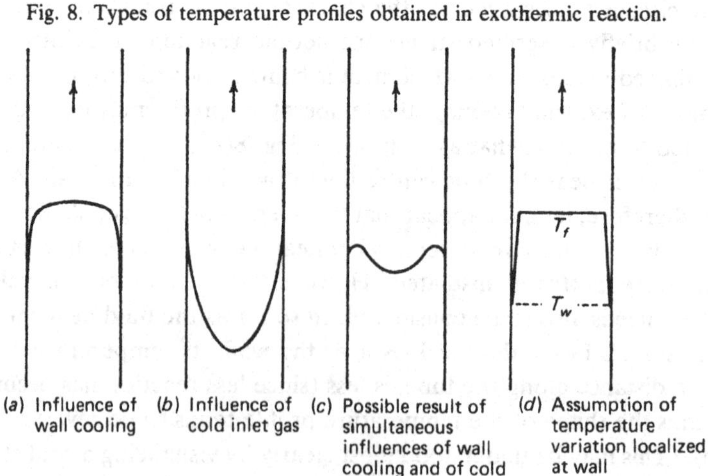

but the difficulties involved in making an *a priori* estimate of the reactor performance are obviously very great. Moreover, there is an additional uncertainty due to the temperature variation within the individual catalyst pellets. As will be shown in a later chapter, this may amount to 50 °C or more. Relative to these two sources of error, the other factors, such as diffusion, which cause deviation from the plug-flow assumption are often much less important [**2**].

The temperature profile and its effect on reactor performance have to be calculated for any particular case (e.g. [**3**]). The following example, though only approximate, will show how the important parameters enter the problem.

***Example* 3.8**

By considering the radial conduction of heat, and the effect of temperature upon reaction rate, show that the design of a tubular reactor of radius a, in which the wall is maintained at a fixed temperature, T, will not be significantly affected by temperature variations within the reactor if

$$\frac{E_A}{RT^2} \cdot \frac{r\Delta H a^2}{\lambda} \ll 1.$$

Here r is the reaction rate at temperature T, E_A is the energy of activation, ΔH the heat of reaction, and λ the thermal conductivity within the reactor.

For the case of non-adiabatic and jacketed reactors the easiest method of making some rough allowance for the temperature variation is to assume that it is localized at the wall. That is to say it is assumed that the temperature of the system has a constant value T_f over the cross-section of the reacting fluid (as in the P.F.A.) but that it changes abruptly and stepwise to a value T_w at the wall (Fig. 8(d)). This procedure is sometimes called 'lumping', whereby the whole problem of heat transfer within and from a packed bed to the wall is 'lumped' into a single, apparent, heat-transfer coefficient.

This is necessarily only a very rough approximation but it represents a significant advance on the assumption that T_f and T_w are equal. On this basis the calculation is carried through on similar lines to the use of the P.F.A. for the adiabatic reactor (as described in § 3.2 above and also in Appendix II to this chapter) except that a wall heat-transfer term is now included in the energy balance. To be more explicit, consider a cylindrical tubular reactor and let dA be the wall area corresponding to the element of reactor volume dV_r already shown in Fig. 6. If r is the tube radius it is readily seen that $dA = 2dV_r/r$. Hence the amount of heat transferred from the fluid to the wall in the element dV_r is

$$\frac{2U}{r}(T_f - T_w)\,dV_r, \text{Js}^{-1},$$

where U is the heat-transfer coefficient.

An energy balance equation can now be set up, enabling the temperature change dT_f (or rather ΔT_f over a small but finite increment of V_r) to be calculated. This equation is similar to that described in Appendix II except that it now includes the above heat-transfer term and also it applies only to the reactor element and not to the whole length of the reactor between the inlet and a given cross-section. For this reason it has to be solved, using standard trial-and-error methods, at the same time as the mass balance equation of the type of equation (3.1) is also solved. For this purpose one first guesses a value of ΔT_f over the element ΔV_r; then, using the mean value of T_f in the element, the amount of reaction occurring within its volume is evaluated. Finally, this estimated amount of reaction is inserted in the differential energy balance equation in order to determine whether the originally guessed value of ΔT_f was a self-consistent one. The method actually involves a double trial-and-error procedure in so far as it is assumed that the radius r is already known. If, after carrying out a sizing by the method outlined, the length of the reactor comes out to be too small in relation to the initially assumed radius, the whole calculation must be repeated using a more suitable initial choice. As well as this stepwise procedure, [**4**], analytical treatments have been provided, e.g. [**5**].

3.7 Transverse temperature gradients: Baron's method

A more soundly based method of estimating the performance of fixed-bed tubular reactors was first put forward by Baron [**6**]. This method, which has been further developed [**7**], avoids the above assumption of a constant temperature over the cross-section. In fact it provides a means of estimating the temperature profile as well as the output of the reactor.

Before outlining this method an important point needs to be mentioned. This is that any allowance for the transverse variation of temperature must always be accompanied by a corresponding allowance for the transverse diffusion of the reagent. A high temperature and a high reaction rate in the region of the tube centre results in a rapid depletion of the quantities of reagents in this region. Therefore it gives rise to steep transverse concentration gradients. The result is an inward radial diffusion flow of reagent and a corresponding outward radial diffusion flow of reaction product. But for the existence of these diffusion flows between the central regions and the peripheral ones, the former regions would become almost inactive, owing to the reagent concentration falling towards zero at some little distance beyond the reactor inlet. This is indicated diagrammatically in Fig. 9. Obviously enough, any design method which allowed for the transverse variation of temperature without simultaneously allowing for the transverse diffusion, would result in the estimated size of reactor being unnecessarily large.

One of the basic assumptions of this treatment is that there is uniformity of the *mass* flow rate* over the cross-section. This is the only remaining vestige of the original plug-flow assumption.

Consider an element of reactor volume in the shape of a rectangular block of sides dx, dy and dz, the z direction being the direction of flow

Fig. 9. Tendency in a tubular reactor for the occurrence of a region depleted of reagents.

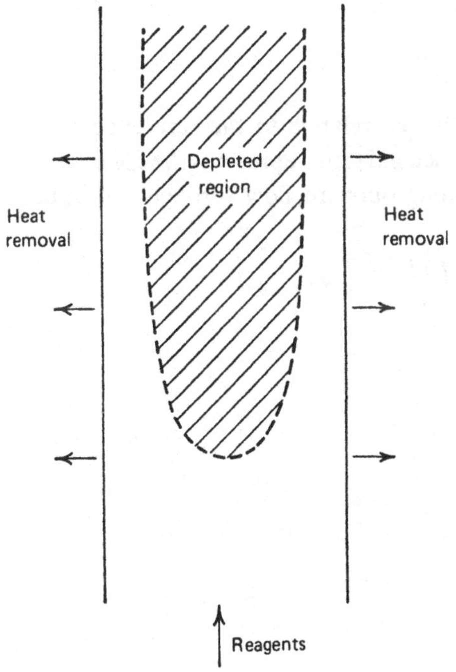

Depleted region
Heat removal
Heat removal
Reagents

* Constancy of *linear* velocity would appear to have been a rather sounder assumption on physical grounds as applied to turbulent flow. (It should be noted that when the temperature, and therefore the density, varies, mass velocity and linear velocity do not bear a constant ratio to each other.)

as in Fig. 10. It will be assumed that the fluid and catalyst temperatures are equal within the element and also that there is negligible temperature variation within the interior of the catalyst particles. Strictly speaking a further assumption, which is implicit in the use of the infinitesimal calculus, is the supposition that the catalyst can be considered as if it were a continuous medium. In fact, of course, it exists as discrete particles. However, the error in using the calculus is probably not very large provided that, over a distance the size of a granule, the temperature and concentration do not change very much in comparison with their absolute magnitudes.

Let C be the heat capacity at constant pressure per unit mass of the reacting fluid and let λ be the effective thermal conductivity of the contents (fluid plus catalyst) of the element as a whole. For simplicity these quantities will be supposed to be temperature-independent. Then if g is the mass flow rate through unit area of the (x, y)-plane and if T is the temperature (relative to a reference temperature) at the bottom face of the element, we have for the net flow of energy through this face

$$CgT\,dx\,dy-\lambda\frac{\partial T}{\partial z}\,dx\,dy.$$

The first term represents the enthalpy, relative to the reference state, which is carried by the fluid and the second term represents the conduction through the face. The corresponding outward flow through the upper face of the rectangular block is

$$Cg\left(T+\frac{\partial T}{dz}\,dz\right)dx\,dy-\lambda\left(\frac{\partial T}{\partial z}+\frac{\partial^2 T}{\partial z^2}\,dz\right)dx\,dy.$$

Fig. 10. Element of reactor volume.

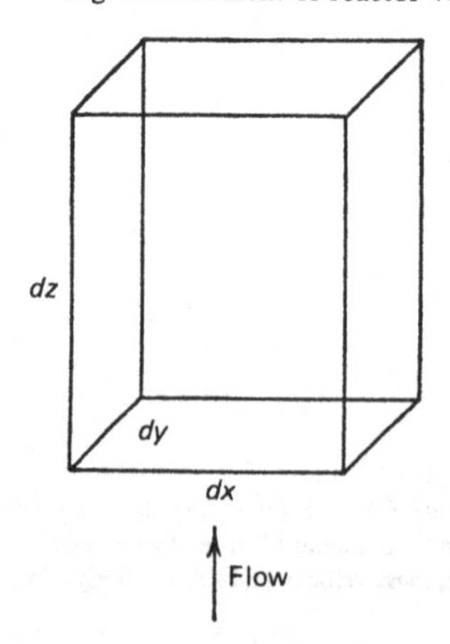

Similar expressions can be written for the energy flows through the remaining four faces, except that for those the terms containing g vanish.

When the system is at a steady state, as is here assumed, the net outflow of energy from the element must be equal to the rate at which heat is generated by reaction. Thus if r' is the reaction rate per unit mass of catalyst whose bulk density is ρ_c, and if ΔH is the enthalpy change in the reaction, we have

$$Cg\frac{\partial T}{\partial z}\,dx\,dy\,dz-\lambda\left(\frac{\partial^2 T}{\partial x^2}+\frac{\partial^2 T}{\partial y^2}+\frac{\partial^2 T}{\partial z^2}\right)dx\,dy\,dz$$
$$=-r'\rho_c\Delta H\,dx\,dy\,dz.$$

or

$$Cg\frac{\partial T}{\partial z}-\lambda\left(\frac{\partial^2 T}{\partial x^2}+\frac{\partial^2 T}{\partial y^2}+\frac{\partial^2 T}{\partial z^2}\right)=-r'\rho_c\Delta H. \tag{3.14}$$

Similarly, if c_i is the concentration of species i, D_i is its effective diffusion coefficient,* u is the apparent linear velocity of flow (i.e. the volume flow rate through any cross-section of the reactor divided by the *total* area of this cross-section and including the area occupied by the catalyst), we have for the mass balance of each species in the element an equation of the type

$$\frac{\partial(uc_i)}{\partial z}-\frac{\partial}{\partial x}\left(D_i\frac{\partial c_i}{\partial x}\right)-\frac{\partial}{\partial y}\left(D_i\frac{\partial c_i}{\partial y}\right)-\frac{\partial}{\partial z}\left(D_i\frac{\partial c_i}{\partial z}\right)=r'\rho_c. \tag{3.15}$$

In this equation r' is taken as positive for substances formed in the reaction and negative for substances which are destroyed. For simplicity it has been supposed that all stoichiometric coefficients are unity; if not, they would need to be introduced appropriately. It has also been assumed implicitly that λ and D are isotropic, i.e. they have equal values in the x, y and z directions.

Baron and Smith [7] proceeded to simplify the equations by neglecting diffusion and conduction in the direction of flow† and also by use of the cylindrical coordinates r and z, which are appropriate to the kind of symmetry existing in most forms of tubular reactor. They also used the previously mentioned assumption concerning the constancy of the mass flow rate.

The practical solution of the equations then turns on putting them into appropriate finite difference forms and solving them numerically from given initial values corresponding to the known conditions at the reactor inlet. The process may be regarded as a double stepwise integration; in

* The significance of D_i will be discussed in Chapter 5.

† See Chapter 5.

a small element of reactor length, starting at the inlet, the equations are first integrated stepwise over the radius; the procedure is then repeated for a second element of length and so on.

This procedure poses, in principle, few problems for a computer, but experimental data on, for example λ, the D_i, and the temperature dependence of k, are necessary for the computation to be carried out. Baron and Smith applied the method to the oxidation of sulphur dioxide and obtained estimated degrees of conversion rather lower than were observed experimentally; but yet not a great deal lower – only some 20%. This is probably as good as could be expected in view of uncertainties in the data and approximations in the method. In another example [**8**], the temperature and composition profiles were calculated for a tubular reactor for the vapour-phase catalytic chlorination of benzene.

In principle, the design of fixed-bed tubular reactors is a computational problem in which variations of flow, temperature, and composition can all be allowed for. However, doubts expressed by Van Heerden [**2**] remain true; the experimental information necessary for such a comprehensive fixed-bed reactor design calculation to be worthwhile is rarely likely to be available. This remark particularly concerns the calculation of the temperature distribution in view of its very sensitive effect on the reaction rate. Uncertainties in the knowledge of the thermal conductivity of the bed, and of its temperature coefficient, can undoubtedly have a large effect. Also a sufficiently comprehensive theory should be capable of allowing for the temperature difference between the catalyst and the reacting fluid and also for the temperature variation within the catalyst itself, which may be 50 °C or more. Further discussion on these points concerning the temperature will be found in Chapter 7.

3.8 Pressure drop

In making estimates of reactor volume by use of the P.F.A. it is often assumed that the pressure is constant over the length of the reactor. In fact it must vary appreciably and especially in reactors containing solid catalyst. Methods of estimating pressure drop in packed beds are available in the literature, e.g. [**1, 9**]. Using these methods the mean of the inlet and outlet pressures can be estimated and it will often be satisfactory to assume that the whole of the reactor operates at this mean. However, for gas reactions at pressures below, say, 10 atmospheres, where the pressure drop through the reactor may be significant in comparison with the mean pressure, a stepwise procedure should be adopted. Here the pressure change is estimated over small increments of reactor length, and simultaneous allowance can also in principle be made for changes in fluid density and viscosity due to the reaction.

3.9 Summary

It may be said that whenever the temperature of the reacting fluid can be held approximately constant over the cross-section, errors from other sources (e.g. velocity gradients and dispersion, for which see Chapter 5) will be acceptable in using the plug-flow assumption. Two rather unusual examples of continuous tubular reactors in which plug flow *is* an adequate approximation can here be mentioned. The first arises in continuous brewing at what is called the 'mashing' stage. A constant-temperature reaction between milled malt and hot water is carried out by passing the mixture through a coiled and jacketed tube. Fig. 11 shows such a plant which can process 1500 kg h^{-1} of malt. The mixture of water and swollen malt does indeed pass through in 'plug' flow.

The second example arises in the oxidation of the lower paraffins. This oxidation proceeds via intermediates such as acetaldehyde, formaldehyde and methanol. Complex chain reactions are involved; the reaction tends to go either not at all, or to complete combustion. Given precise temperature control at entry, a tubular reactor can be used to provide the desired intermediates at high yield. A carefully controlled flow rate ensures that the optimum yield obtains in the stream leaving the reactor and passing to a quench system where further reaction, leading to degradation of the desired intermediates, is prevented. The tubular reactor is said to be rather less than a metre in diameter and a hundred metres or more in length.

However, whenever the temperature does vary significantly in a transverse direction the deviations from the predictions of the elementary design method of § 3.2 can be very great indeed, owing to the large temperature dependence of the velocity constant for most reactions.

This source of error is particularly significant in the case of fixed-bed catalytic reactors having transverse heat removal through the walls. Theoretical methods of dealing with this problem have been put forward, but to apply and test these methods requires far more experimental information than is usually available. The deterioration of the catalyst is an additional factor causing uncertainty.

Because of these difficulties the actual industrial design of fixed-bed tubular reactors is often carried out by purely empirical methods based on the discovery of satisfactory operating conditions in a pilot scale plant. One such method is 'stepping up'. If it is found that a pilot scale reactor in the form of a tube x cm in diameter and y cm long gives satisfactory performance, the method assumes that a hundred such tubes put together in parallel as a full-scale reactor will be equally satisfactory. But x and y must remain the same because of the difficulty, or the impossibility, of achieving dimensional similarity

Fig. 11. A tubular reactor for continuous 'mashing'. The photograph shows the return bends of the long tube, and is by courtesy of The A.P.V. Co. PLC, and Horlicks PLC.

between tubes of different sizes whenever there is heat transfer as well as reaction. This is because heat transfer, which depends on *area*, and chemical reaction, which depends on *volume*, do not change in the same ratio when there is an alteration of diameter.

Appendix I Isothermal reaction with plug flow

The following problem will be solved for the purpose of showing how equation (3.5) may be integrated.

The gas leaving an ammonia oxidation plant is cooled rapidly to atmospheric temperature in order to remove the bulk of the water vapour and it then contains 9% (molar) of nitric oxide, 1% of nitrogen peroxide and 8% of oxygen. Before entering absorption towers for the production of nitric acid the gas is allowed to oxidize until the NO_2:NO ratio reaches 5:1.* It is required to calculate the volume of a tubular reaction vessel which would suffice for this purpose if the cooling is assumed to be efficient enough for the temperature of the reacting gas to remain constant at 20 °C. The gas flow rate at the reactor inlet is 10 000 $m^3\,h^{-1}$ (measured at 0 °C and 1 atm) and the gas pressure is 1 atm.

The reaction

$$2NO+O_2=2NO_2$$

is known to be a homogeneous gas reaction which is virtually irreversible and whose order corresponds to its stoichiometry (see the remarks made in Chapter 2). Its rate is therefore proportional to the term $[NO]^2[O_2]$. When these concentrations are expressed in kmol per m^3 and the time in seconds, the work of Bodenstein and Lindner [**10**]† shows that the velocity constant at 20 °C has the value of 1.4×10^4.

The design equation (3.6) now takes the form

$$V_r=G\int_{y_i}^{y_e}\frac{dy}{k[NO]^2[O_2]}, \tag{3.16}$$

where for the time being k will be kept within the integral sign, in order to obtain an equation suitable also for the purposes of Appendix II.

* The necessity for a relatively high NO_2:NO ratio arises from the fact that the absorption reaction: $3NO_2+H_2O=2HNO_3+NO$ is very reversible.

† In Appendices I and II the Bodenstein and Lindner values have been re-expressed by using seconds as the unit of time and have also been divided by 2. This is because Bodenstein and Lindner defined their velocity constants with respect to *half* concentrations of NO and NO_2. We here take NO_2 (or NO) rather than O_2, as being the 'key component' (cf. § 2.4).

It remains to express the quantities y, [NO] and [O_2] in terms of a single variable.

In order to obtain the simplest illustration we shall here neglect the existence of a little N_2O_4 in equilibrium with the NO_2. (N_2O_3 is also neglected and even more justifiably.) Consider 1 kmol of gas entering the reactor, and let x be the fraction of NO in the gas which has been converted to NO_2 (so $x=0.1$ in the feed gas, which contains 9% NO and 1% NO_2). We can work out the composition that the gas would have had for $x=0$ from the stoichiometry, as well as the composition for any other value of x, as follows:

	kmol at inlet	kmol $x=0$	kmol x
NO	0.09	0.100	$0.100\,(1-x)$
NO_2	0.01	0	$0.100\,x$
O_2	0.08	0.085	$0.085-0.5\,x$
N_2	0.82	0.82	0.82
Total	1.00	1.005	$1.005-0.05\,x$

In this case it is not possible to use equations such as (3.7) and (3.8) directly, because of the presence of excess oxygen and diluent nitrogen, but the following procedure is in fact similar.

At any cross-section where the conversion is x, the mole fractions of NO and O_2 are

$$\frac{0.100-x}{1.005-\frac{1}{2}x}, \quad \text{and} \quad \frac{0.085-\frac{1}{2}x}{1.005-\frac{1}{2}x} \tag{3.17}$$

respectively. The total molar gas concentration, assuming ideal gas behaviour, is P/RT kmol per m^3, where P is the pressure, T the temperature and R the gas constant, in appropriate units.

Now considering equation (3.2), we can replace G and dy in equation (3.16) by m_f and dx, where m_f is the molar feed rate of ($NO+NO_2$). Substituting for the concentrations, [NO] and [O_2], in terms of their mole fractions, from equation (3.17), times the total gas concentration, P/RT, we obtain from equation (3.16)

$$V_r = m_f \int_{x_{in}}^{x_{out}} \frac{R^3T^3}{P^3k} \frac{(1.005-0.05x)^3}{0.01(1-x)^2(0.085-0.05x)}\,dx. \tag{3.18}$$

If the gas pressure, P, is 1 atm, and T is measured in K, then $R=0.0821$. The molar feed rate of ($NO+NO_2$), m_f, is obtained from the inlet volume flow rate of the gas, of which ($NO+NO_2$) comprises 10%. 10 000 m^3 h^{-1} of gas, measured at 0 °C and 1 atm, is $10\,000/(3600\times 22.41)$ kmol s^{-1}. Thus $m_f=0.1\times 10\,000/(3600\times 22.41)=0.0124$ kmol s^{-1}.

Substituting in equation (3.18), we have

$$V_r = 6.86 \times 10^{-4} \int_{x_{in}}^{x_{out}} \frac{T^3}{P^3 k} \frac{(1.005-0.05x)^3}{(1-x)^2(0.085-0.05x)} dx. \quad (3.19)$$

For this problem $x_{in} = 0.1$, and x_{out} is such as gives a 5:1 ratio of NO_2:NO. Clearly $x_{out} = 5/6 = 0.833$. $P = 1$ atm, and $T = 293.2$ K. The integral can be evaluated graphically or numerically, and when this is done V_r is found to be 110 m^3.

Hence the tubular reactor must have a volume of 110 m^3. When asked, 'What is the next stage in the design beyond this point?' some students reply unthinkingly that it is to determine the time of passage which is necessary for the reaction to take place to the required extent! But of course this is essentially what has been done already. The figure of 110 m^3 is the volume of reactor which gives the required time of passage at the given flow rate. The next stage in the design, which now goes beyond the scope of chemical kinetics, is to choose the length:diameter ratio. What are the engineering factors which affect this choice?

It is of interest to inquire how large would have been the error if, instead of assuming, as above, that each element of fluid remains at constant pressure as it flows along the reactor, we assume that it remains at constant volume. If this were the case we could write the equation for the chemical kinetics in the form

$$\frac{d[NO_2]}{dt} = k[NO]^2[O_2].$$

After allowing for the stoichiometry, this equation can be integrated directly. The result is to show that a time of 39.4 s will result in the NO_2:NO ratio reaching the required value. The flow rate being 10 000 $m^3\,h^{-1}$, this residence time corresponds to a reactor volume of 117 m^3. In this instance, therefore, the error due to neglecting the change of volume is rather small and this is because of the diluteness of the reacting components. In other situations, e.g. § 3.3, the error can be very considerable. The student should ask himself, however, why the reactor volume as thus calculated comes out greater and not less than the more accurate value of 110 m^3 as previously estimated.

Appendix II Adiabatic reaction with plug flow

We shall here work out the volume of a tubular reactor suitable for dealing with the same gas as in Appendix I, and resulting in the same degree of conversion, but operating adiabatically instead of isothermally. The inlet temperature will be taken as 20 °C, as before, and the calculation will again be based on the use of the P.F.A.

For the purposes of this problem, equation (3.19) remains entirely applicable, but it is no longer possible to take k and T as constant. Instead it is necessary to work out values of T as a function of the variable x, by means of an energy balance, and subsequently to tabulate values of k for suitable corresponding values of x and T. With this information the equation can be integrated graphically or by computer.

The change of heat of reaction with temperature will be allowed for in this illustration, but for reasons of simplicity we shall take the heat capacities to be constant. The required data are as follows in units of kJ, kmol and °C.

	ΔH^f_{298}	c_p
NO	90 200	29.8
NO_2	33 800	37.9
O_2	—	29.3
N_2	—	29.1

The heat of reaction at 25 °C is therefore 56 400 kJ per kmol of NO_2. Using the specific heat data, the heat of reaction at 20 °C is 56 367 kJ per kmol of NO_2.

As before, we consider 1 kmol of gas entering the reactor, which contains 0.1 mol of $(NO+NO_2)$. The energy balance between the reactor inlet, where the temperature is 20 °C, and any given cross-section, where the conversion of NO is x, can be calculated by taking the reaction $(x-x_{in})$ at 20 °C, and then using the heat so released to heat up the gas mixture, of conversion x, from 20 °C to temperature t.

Thus

$$56\,367\times 0.1(x-x_{in})=(t-20)\,[29.8\times 0.100(1-x)+37.9\times 0.100x$$
$$+29.3\times(0.085-0.05x)+29.1\times 0.82].$$

We also note that the conversion of NO at entry to the reactor, x_{in}, is 0.1. Hence

$$t=20+\frac{5637x-563.7}{29.33-0.655x}$$

We can thus obtain values of t for chosen values of x up to $x_{out}=0.833$, which is, as in Appendix I, the desired outlet conversion. A table of such data is given below, together with the corresponding values of k, as taken from the data of Bodenstein and Lindner.

x	0.0	0.01	0.02	0.03	0.04	0.05	0.06	0.07	0.0733
t(°C)	20	39	59	78	98	117	137	157	164
$k\times 10^{-4}$	1.40	1.30	1.12	0.97	0.87	0.81	0.77	0.73	0.72

Having these results, all the information is now available for working out values of the integrand in equation (3.23) and thereby to perform a graphical integration by plotting the integrand against x. The result is

$$V_r = 550\ \text{m}^3.$$

It will be seen that this is some five times larger than is required for the reactor of Appendix I which was specified for isothermal operation. This large increase is due to two effects which operate in the same direction: (1) temperature rise causes a reduction in gas density and thereby causes the concentrations [NO] and [O_2] to diminish; (2) the oxidation of nitric oxide is a very exceptional reaction in so far as its velocity constant becomes smaller with rise of temperature.

In the case of most exothermic reactions having the normal positive temperature coefficient of reaction rate, adiabatic conditions would result in a *smaller* necessary reactor volume than for isothermal conditions. This statement applies, however, only to reactions which are not appreciably reversible. It may easily occur with an exothermic reversible reaction that a desired conversion is unattainable without cooling, since the reaction mixture would otherwise heat up adiabatically to a temperature at which further conversion were impossible because equilibrium had been reached.

The following examples illustrate the principles discussed in this chapter.

Example 3.9

Pure gaseous A at 2.5 bar is fed at a rate of 4.0 kmol h^{-1} to a plug-flow reactor where it reacts reversibly and isothermally at 609.5 K with elementary kinetics

$$A \underset{k_2}{\overset{k_1}{\rightleftharpoons}} 2B$$

where k_1, based on the rate of destruction of A, = 170 h^{-1} and the equilibrium constant, K_p, = 2.5 bar.

Use a numerical or an analytical method to estimate the size of reactor needed for 30% conversion.

Note: $\displaystyle\int \frac{dx}{a^2 - x^2} = \frac{1}{2a} \ln \frac{(a+x)}{(a-x)}$ for $x^2 < a^2$.

[*Answer.* 0.236 m^3.]

Example 3.10

A packed-bed catalytic reactor is to be used to decompose a gas A according to the equation $A \rightarrow 2B$. The reaction rate may be written

$$\text{Rate (kmol h}^{-1}(\text{kg catalyst})^{-1} = \frac{0.085 p_A}{1 + 0.15 p_A}$$

where p_A is the partial pressure of A in bar.

The feed is a 2:1 mixture of A and an inert gas, respectively, and is supplied at 5 bar at a rate of 2.2 kmol h^{-1}. The catalyst pellets have a relative density of 2.4 and the packed-bed voidage is 0.4.

If the reactor is isothermal and plug flow can be assumed, find what reactor volume is required to convert 80% of the entering A to B.

[*Answer.* 9.16×10^{-3} m^3.]

Example 3.11

The reaction $A \rightarrow B$ will be carried out in a tubular reactor. The reaction is autocatalytic, the rate being given by $k[A][B]$. The effluent from the reactor is divided into two streams. One goes forward for product purification; the other, which is R times as large, is recycled to mix with the feed stream, which is a solution, concentration a_0, of A in an inert solvent, before entering the reactor.

The desired conversion of A is to be 90%. It can be shown that the minimum reactor volume for this conversion will be required with $R = 0.43$. How much larger will the reactor have to be if $R = 0.2$?

[*Answer.* 1.035 times as large.]

Example 3.12

The reaction $2A \rightleftharpoons B + C$ is to be carried out in an isothermal plug-flow tubular reactor. The feed flow rate is 10 m^3 h^{-1}. The initial concentration of A is 5 kmol m^{-3}, and the feed contains no B or C. The reaction kinetics are given by

rate of production of $B = k_1 C_A^2 - k_2 C_B C_C$,

where C_A, C_B, C_C are the concentrations in kmol m^{-3} of A, B and C, and $k_1 = 7.5$ m^3 $kmol^{-1}$ h^{-1} and $k_2 = 0.5$ m^3 $kmol^{-1}$ h^{-1}.

What reactor volume is required if the final concentrations of B and C are to be 98% of their equilibrium values?

[*Answer.* 1.34 m^3.]

Symbols

Symbols defined in, and referring to, particular problems are not included. The context should give meaning to subscripts not shown below.

- c Concentration, kmol m^{-3}.
- C Heat capacity per unit mass, J kg^{-1} K^{-1}.
- D (Effective) diffusion coefficient, m^2 s^{-1}.
- g Mass flow per unit cross-section of reactor, kg m^{-2} s^{-1}.

G	Mass flow rate, kg s^{-1}.
ΔH	Enthalpy change of reaction, J $kmol^{-1}$ (stoichiometric).
k	Velocity constant of reaction, units depend on kinetics.
m_f	Molar flow rate of feed, kmol s^{-1}.
P	Pressure, bar
R	Gas constant, bar m^3 $kmol^{-1}$ K^{-1}.
r	Reaction rate, kmol m^{-3} s^{-1}.
r'	Reaction rate, kmol s^{-1} (kg catalyst)$^{-1}$.
t	Time, s.
T	Absolute temperature, K.
u	Superficial velocity, m^3 s^{-1} per m^2 reactor cross-section.
U	Heat transfer coefficient, J m^{-2} s^{-1} K^{-1}.
v	Flow rate in reactor, m^3 s^{-1}.
V_r	Reactor volume, m^3.
W_r	Mass of catalyst in reactor, kg.
x	Fractional conversion of feed.
y	kmol of product per unit mass of fluid.
α, β	Reaction orders.
ε	see equation (3.7).
λ	Effective thermal conductivity of reactor contents, J m^{-1} s^{-1} K^{-1}.
ρ_c	Bulk density of catalyst, kg per m^3 of reactor volume.
τ	Space-time, s.

References

1. Hougen, O., Watson, K. M. and Ragatz, R. A., *Chemical Process Principles*, part III (Wiley & Co., 1959).
2. Van Heerden, C., *Chem. Engng Sci.*, 1961, **14**, 101.
3. Schoenemann, K. and Hofmann, H., *Chem.-Ing.-Tech.*, 1957, **29**, 665.
4. Smith, J. M., *Chemical Engineering Kinetics* (McGraw-Hill, 1956), pp. 160 and 331 ff.
5. Chambré, P. L., *Chem. Engng Sci.*, 1956, **5**, 209.
6. Baron, T., *Chem. Engng Progr.*, 1952, **48**, 118.
7. Smith, J. M., *Chemical Engineering Kinetics*, pp. 346 ff.
8. Kramers, H. and Westerterp, K. R., *Elements of Chemical Reactor Design and Operation* (Netherlands University Press, Amsterdam, 1963).
9. Coulson, J. M. and Richardson, J. F., (ed.), *Chemical Engineering* (3rd edn, Pergamon Press, 1978), Vol. 2, Ch. 4.
10. Bodenstein, M. and Lindner, *Z. phys. Chem.*, 1922, **100**, 87.

4

Continuous stirred tank reactors

4.1 The 'perfect mixing' assumption

The general characteristics of the C.S.T.R. have been noted already in § 1.5. The reactor consists of a well-stirred tank into which there is a steady flow of reacting material and from which the (partially) reacted material passes continuously. It will be seen later often to be advantageous to have several C.S.T.R.s in series. The important difference, as compared to the tubular reactor, is the deliberate attempt to achieve good mixing in each vessel, and this is done in order to ensure that the volume is fully utilized for reaction, with no appreciable dead space. In a tube-shaped reactor the whole of the internal volume is adequately swept through by the reacting fluid, but this would not be the case, in the absence of stirring, in the kind of vessel used in the C.S.T.R.

Whereas in the instance of the tubular reactor the simplest design method is based on the assumption of plug flow, the corresponding simplifying assumption for the C.S.T.R. is that there is complete mixing in each vessel. How 'complete', or 'perfect', does mixing have to be for the vessel to be 'completely mixed'? If the circulation time of a fluid element within the vessel is, say, one hundredth of the average time spent within the vessel, then mixing may well be 'perfect' in the sense of this chapter. Nevertheless the relation between what may be called the *mixing* time and the *circulation* time is complex; neither quantity is very easy to define precisely.

A field where these concepts are of central concern is that of industrial fermenters, and other instances of somewhat viscous mixtures (with or without aeration). With such reactors the cost of the stirring, and of the cooling necessary to remove the heat generated by the stirring, can be a significant part of the total cost of the process. A short review of this field has been given by Bryant [**1**].

With complicated kinetic schemes, involving, for example, consecutive reactions, the 'degree of mixing' requires further consideration. It is found that the proportions of different products can be quite sensitive to the degree of mixing in small regions of the fluid – far smaller than we need be concerned about when dealing with simpler kinetics. This will be referred to again in later chapters.

For the moment, we shall assume that the performance of a C.S.T.R. will differ only negligibly from that calculated on the assumption of 'complete mixing'.

As was noted earlier in § 1.5, the mixing has two important consequences (which are really the same): (1) a bypassing effect; (2) a stepwise change in reagent concentration from one vessel to the next. As regards the first, it will be obvious that the result of good mixing is that a molecule added at the reactor inlet may be present at the next instant almost anywhere* in the vessel, and in particular it may be present in the fluid which at that moment is in process of leaving the vessel. Conversely, there are molecules present in the vessel which have not succeeded in finding their way into the outlet after a very long time indeed. In brief, there is a broad distribution of the residence-times in each vessel and its form will be discussed in the next chapter. For the moment it is sufficient to remark that the existence of the bypassing effect has a beneficial as well as an adverse consequence: on the credit side, the immediacy of the response at the reactor outlet to any change in concentration at the inlet is a very helpful feature in regard to automatic control; on the debit side it is the occurrence of the bypassing which often makes it necessary to carry out the reaction in not one, but in a chain of several C.S.T.R.s.

Only in the case of reactions (e.g. free radical polymerizations) which are extremely rapid relative to the mean residence-time is it possible to avoid excessive loss of reagent when only a single stirred tank is in use.

The other consequence of the mixing is the stepwise change in concentration. If mixing were perfect it would mean that the composition of the reacting fluid would have a uniform value throughout the volume of a particular vessel. It follows that this composition would also be that of the fluid leaving that vessel. Therefore it would differ by a finite amount from the composition of the fluid in the next vessel in the series. That is to say, logical consistency in using the assumption of perfect mixing requires us to suppose that the fluid entering a vessel is instantaneously mixed into the fluid already present and that the time during

* Quite literally 'anywhere' in the limiting case of perfect mixing, and with equal probability.

which the new material passes through intermediate concentrations is effectively zero.

For most types of kinetics the stepwise change in concentration results in the average reaction rate being much smaller than it would be if the same feed materials were allowed to react batchwise or in a tubular reactor. Therefore, to obtain the same output, the volume of reaction space must be larger – and in some cases very much larger – and to a greater degree the smaller the number of vessels in series.*

This may be understood in a preliminary way by considering the reaction between two substances A and B, each available in feed solutions whose concentrations are 20 kmol m^{-3}. If equal volumes of these two solutions were brought together in a batch or tubular reaction process, the concentrations of each reagent immediately after mixing would be 10 kmol m^{-3}. Let it be supposed that the economics of the process permitted a reaction time long enough for the concentrations to fall to 0.5 kmol m^{-3} – i.e. 95% conversion. The batch or tubular reaction process would thus correspond to reagent concentrations diminishing along the curve in Fig. 12. By contrast, if the same reaction using the same feed solutions were carried out in a *single* vessel C.S.T.R., and if it were required to achieve the same degree of conversion, the concentrations *throughout the volume* of the tank would necessarily be 0.5 kmol m^{-3} (assuming perfect mixing). That is to say, the stationary concentrations in the tank would be the same as the concentrations *at the termination* of the batch or tubular reaction process and would correspond to point P on the figure. Obviously the reaction rate would be much reduced and in fact would be measured by the tangent at P and this is the smallest tangent over the whole curve as far as point P.

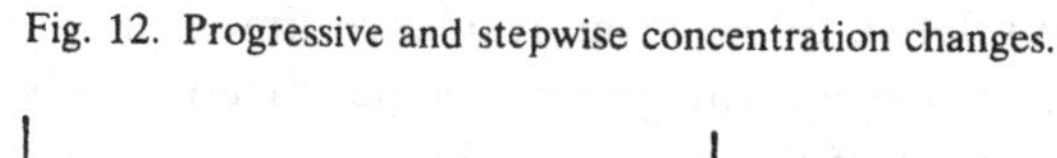
Fig. 12. Progressive and stepwise concentration changes.

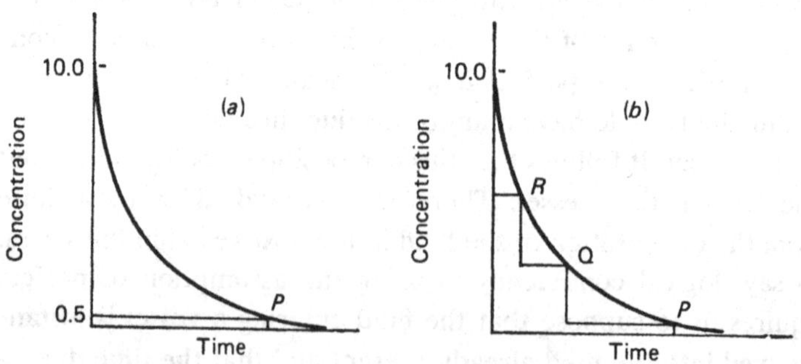

* This is not necessarily the case, however, if the reaction is regarded as taking place adiabatically, rather than isothermally. Why is this?

The necessary increase in reactor volume, which is the consequence of this reduced reaction rate, can obviously be diminished by using several tanks in series, thus obtaining a stepwise fall in concentration as is indicated for the case of three tanks in Fig. 12. Nevertheless, the volume per unit of output will still be greater than would be needed for a batchwise reactor (apart from the effect of emptying and recharging time) or for a tubular reactor having plug flow.

The extra capital cost of the C.S.T.R. arising from its greater volume is not often very significant. Such reactors are most frequently used for liquid-phase reactions taking place at atmospheric pressure, and for such processes the cost of providing volumetric capacity is usually only a trivial part of the total. Nevertheless, it is of the utmost importance that the plant designer should be aware of the fact that a greater volume *is* needed. A mere scaling up from a batchwise reaction in the laboratory might result in a plant which gave only a fraction of the required output. Moreover, it will be an aspect of the designer's art to discover that number of C.S.T.R.s in series which will minimize the reactor cost.

An interesting reactor which is essentially a C.S.T.R. was used to produce town gas by the hydrogenation of a light petroleum vapour [**2**]. One of the necessities of the process is to obtain a temperature high enough for reaction to set in readily, but not so high that side reactions take place. If the reacting gas mixture is made to circulate rapidly within the reactor the incoming gases are quickly brought up to reaction temperature, but the development of hot spots is avoided because of the dilution effect.

4.2 Calculation of reactor volume

The theory of the C.S.T.R., in which we assume perfect mixing, is a straightforward application of the material balance equation, equation (2.1), and the reaction rate expression, equation (2.2). For simplicity consider the liquid-phase first-order decomposition of a reagent A. As we saw in Chapter 2, a perfunctory and erroneous consideration of the problem would be to write: $dc_A/dt = -kc_A$, followed by: $c_A = c_A^0 e^{-kt}$. This might lead to the conclusion that the product composition, c_A, is given by

$$c_A = c_A^0 e^{-kV/v}, \tag{4.1}$$

where c_A^0 is the feed concentration of A, V is the voume of the C.S.T.R., and v is the volume flow rate through it.

This is incorrect, as pointed out in Chapter 2; a correct application of the balance equation (2.1) gives us

$$vc_A^0 = vc_A + rV. \tag{4.2}$$

Since the C.S.T.R. is well stirred, we can integrate the rate over all elements dV, giving rV (because r is constant throughout V). Substituting $r = kc_A$, we obtain

$$c_A = c_A^0/(1 + kV/v), \tag{4.3}$$

which is actually simpler than (4.1). This applies to a C.S.T.R. in the steady state.

We can note immediately that for a given value of kV/v, c_A is higher in equation (4.3) than in equation (4.1). For such a system, equation (4.1) gives the correct answer if a P.F.R. of the same volume is used. This comparative disadvantage of a C.S.T.R. over a P.F.R. is merely another illustration of the problem considered in the previous section and in Fig. 12.

***Example* 4.1**

A C.S.T.R. is used to decompose a dilute solution of A. The decomposition is irreversible and first order, with velocity constant 3.45 h^{-1}. The reactor volume is 10 m^3. What flow rate of feed solution can be treated by this reactor if 95% decomposition is required?

[*Answer.* 1.82 $m^3\,h^{-1}$.]

Let us now consider the liquid-phase reaction

$$A + B \rightarrow X.$$

This will be assumed irreversible and second order, i.e. first order with respect to each of the substances A and B. The reaction rate expression, equation (2.2), becomes

$$r = kc_Ac_B,$$

where c_A and c_B are uniform for the whole tank at any time, since the tank is 'well-mixed'. It follows that the amount of A or B destroyed by reaction in the whole tank is kVc_Ac_B kg moles per unit time. We assume that the total liquid hold-up in the tank is steady and that the flow rate through the tank is constant and equal to v m^3 per unit time.

Application of the mass balance equation, (2.1), to this situation gives us

$$\underset{A\text{ in}}{vc_{A0}} = \underset{A\text{ out}}{vc_A} + \underset{A\text{ destroyed}}{kVc_Ac_B} + \underset{\text{Rate of change of }A\text{ in tank}}{V\,dc_A/dt}. \tag{4.4}$$

Here c_{A0} is the concentration of A in the entering feed stream, and c_A will vary with time if the reactor has not yet reached the steady state. There is a similar equation for B. Note that v is the total flow rate through the tank. If two separate feed streams enter the tank, then v is the sum of the two, and c_{A0} is the concentration of A which would occur if the two feed streams were mixed just before entry to the tank.

Since there is a stoichiometric relationship between c_B, c_{B0}, c_{A0} and c_A, equation (4.4) can in principle be solved to calculate c_A as a function of k, V and t. It should be noted that the value of k is determined by the temperature prevailing in the tank. In general, if the heat of reaction is significant, the temperature will be dependent on c_A. It can be calculated by use of an energy balance. We shall defer this calculation to a later chapter; here we shall assume we know the temperature and hence the value of k.

Equation (4.4) becomes much simpler if we assume that the reactor has attained the steady state. We can then find c_A as a function of V, or can regard the calculation as one to provide the value of V for a required degree of conversion.

***Example* 4.2**

The reactor of Example 4.1 is started up by rapidly filling it with feed solution, and thereafter using a steady flow rate into and out of the reactor of $1.82 \text{ m}^3 \text{ h}^{-1}$. How long will it be before the concentration of A in the outlet stream falls to 6% of that in the feed? (Solve the non-steady-state analogue of equation (4.2).)

[*Answer.* 1.25 h.]

***Example* 4.3**

Ethyl acetate solution of normality 1.21×10^{-2} and sodium hydroxide solution of normality 4.62×10^{-2} are fed at the rates of $11.2 \text{ m}^3 \text{ h}^{-1}$ and $11.3 \text{ m}^3 \text{ h}^{-1}$ respectively into a continuous stirred tank reactor. The latter consists of a single vessel only, and contains 6.0 m^3 of liquor which is in process of reaction. Given that the second-order velocity constant of the hydrolysis reaction is $11.0 \times 10^{-2} \text{ m}^3 \text{ kmol}^{-1} \text{ s}^{-1}$, calculate the normality of the ethyl acetate in the solution leaving the vessel, at the steady state. Also calculate the percentage of the ethyl acetate that has been hydrolysed.

[*Answers.* 2.0×10^{-3}; 67%.]

In general, the higher the kinetic order of a reaction, the smaller the conversion in a C.S.T.R. in comparison with that in a P.F.R. (if both are isothermal). Fig. 13 illustrates this for the decomposition of a single reagent. For a first-order reaction, the feed concentration, c_f, is reduced to a lower value, c_{out}, in a batch reactor than in a C.S.T.R., given equal values of the reaction parameter, kt for the batch reactor, and kV/v for the C.S.T.R. If the reaction is second-order, the conversion is even lower (c_{out} is higher) in a C.S.T.R., at comparable values of the reaction parameter, which is kVc_f/v in this case.

The most common use of C.S.T.R.s is for liquid-phase reactions, for which the necessary reactor volume is likely to be smaller than for a gas-phase reaction, in view of the higher density of liquids. It is interesting, however, to see what effect the change of density of a gas process stream has upon the design calculation. For this purpose we shall look again at the decomposition of acetaldehyde example (see § 3.3).

Fig. 13. Product composition versus reaction parameter.

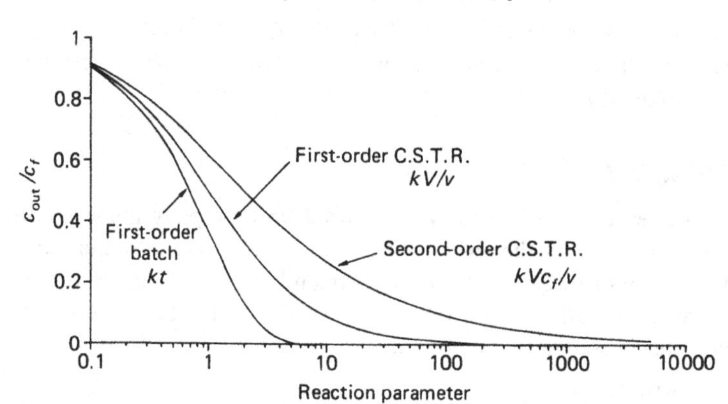

***Example* 4.4**

Acetaldehyde is to be decomposed at 520 °C and 1 atm in a C.S.T.R. The reaction is irreversible and second order with respect to CH_3CHO. The velocity constant at 520 °C is 0.43 $m^3\ kmol^{-1}\ s^{-1}$. The feed rate is 0.1 kg s^{-1} of pure acetaldehyde. What size reactor will be needed (*a*) for 35% decomposition, and (*b*) for 90% decomposition?

Solution

We apply the mass balance equation, equation (2.1), taking care to allow for the change in the number of moles arising from the reaction $CH_3CHO \rightarrow CH_4 + CO$.

We shall use the molar feed rate of acetaldehyde, m_f, and work in terms of x, the conversion of it to $CH_4 + CO$. We obtain

$$m_f = m_f(1 - x_{out}) + rV, \tag{4.5}$$

in which r is based on x_{inside}. Remember that $x_{inside} = x_{out}$ if the vessel is completely mixed. Hence

$$V = \frac{m_f x_{out}}{r}, \quad \text{where } r = k[CH_3CHO]^2.$$

Applying equation (3.7) to this case, we have

$$r = k[CH_3CHO]^2 = k\left[\frac{1-x_{out}}{1+x_{out}}\right]^2\left(\frac{P}{RT}\right)^2 \tag{4.6}$$

Hence

$$V = \frac{m_f}{k}\left(\frac{RT}{P}\right)^2\left(\frac{1+x_{out}}{1-x_{out}}\right)^2 x_{out}. \tag{4.7}$$

This is to be compared with equation (3.9). Substituting the values for the parameters in equation (4.7), we find the following answers:
For 35% decomposition $V = 33.8\ m^3$ (cf. $17.5\ m^3$ for a P.F.R.)
For 90% decomposition $V = 7270\ m^3$ (cf. $619\ m^3$ for a P.F.R.)

***Example* 4.5**

For the two cases in Example 4.4 calculate the space-time (based on inlet gas at 520 °C) and the average residence-time (which is based on outlet gas at 520 °C). Compare the answers with those in § 3.4.
[*Answers.* $\tau = 229$ s, 13.6 h; $\bar{t} = 170$ s, 7.2 h.]

4.3 Stirred tanks in series

The examples in § 4.2 have shown that a C.S.T.R. may be of much greater volume than a tubular reactor, especially if the desired conversion is high. This volume can be reduced by using two or more stirred tanks in series, the volume of each tank being much smaller than for the single C.S.T.R.

Consider Example 4.1. Equation (4.3) shows us that

$$\frac{kV}{v} = \frac{c_{A0}}{c_A} - 1. \tag{4.8}$$

For 95% reaction $kV/v = 19$, and we shall compare this with the figure for a tubular reactor, and for two equal-volume C.S.T.R.s in series. In all cases it will be assumed that 95% of the A fed to the plant is to be decomposed.

For a tubular reactor it is easy to show that equation (2.1) leads to

$$\frac{kV}{v} = \ln(c_{A0}/c_A). \tag{4.9}$$

For 95% reaction this gives $kV/v = 3.0$, which is much less than the value of 19 for a C.S.T.R., and this is what our earlier examples would have led us to expect.

Now consider two C.S.T.R.s in series, each of volume $V_{1/2}$. Equation (4.8) can be used to calculate c_{A1}, the concentration of A in the feed to the second tank. Then we use equation (4.8) again to calculate c_{A2}, the concentration of A in the final product from the second tank. We have

$$\frac{kV_{1/2}}{v}=\frac{c_{A0}}{c_{A1}}-1, \qquad \frac{kV_{1/2}}{v}=\frac{c_{A1}}{c_{A2}}-1. \tag{4.10}$$

We eliminate c_{A1} between these two equations, obtaining

$$c_{A2}=c_{A0}/(1+kV_{1/2}/v)^2 \quad \text{or} \quad \frac{kV}{v}=\frac{2kV_{1/2}}{v}=2\sqrt{\left(\frac{c_{A0}}{c_{A2}}\right)}-2, \tag{4.11}$$

where V is the *total* volume. For 95% decomposition $kV/v=7.7$. This is less than half the volume for a single C.S.T.R., but still over twice the volume of a tubular reactor.

It will be useful to solve the following two examples at this point.

***Example* 4.6**

A first-order irreversible reaction is to be carried out in a chain of n C.S.T.R.s of equal volumes. There is no change of density of the process stream as it passes through the chain. Show that if n is very large, the *total* volume of the chain tends to that of a tubular reactor producing the same final product.

***Example* 4.7**

Two C.S.T.R.s in series are to be used to produce a desired decomposition of a reagent which decomposes according to a first-order law. There is no change of density due to reaction. Show that the minimum total volume is obtained if the two tanks are of equal size.

We saw in Fig. 13 that the higher the kinetic order, the larger does a C.S.T.R. have to be to produce the same conversion as a P.F.R. (and this characteristic is more marked the higher is the desired conversion). For such higher kinetic orders it is even more desirable to use a chain of C.S.T.R.s than for a first-order reaction, since the proportional reduction in total volume is more marked. The total volume is always higher, the higher the order, but less strikingly so if a chain is used.

Fig. 14 shows this comparison. For a first-order decomposition we can use equations (4.8) and (4.11) to plot the curve. For a second-order decomposition of a single reagent we must derive the answers analogous to equations (4.8) and (4.11) to obtain the curve shown. For conversions much higher than 0.8, Fig. 14 is not a convenient representation of the figures, which can, of course, readily be calculated.

In the case of a simple reaction there is no particular advantage in running the tanks in a chain of C.S.T.R.s at different temperatures. For more complex reaction schemes there can however be a significant advantage, as we shall see in a later chapter, but it needs to be appreciable in order to justify the expense of heat exchange equipment required for maintaining the temperature differences in the chain.

Again, Example 4.7 shows that the minimum total volume of a two-tank chain is obtained when the tanks are of equal volume for the case of a first-order reaction. (This applies even if the reaction is reversible, if both forward and backward steps are first order.) This is not true for reactions of other orders. Here the minimum total volume is obtained when the tanks are of unequal size. The optimum ratio of sizes depends upon the kinetics and the desired extent of reaction. It can be shown that for a simple second-order irreversible reaction between reagents fed to the first tank in stoichiometric quantities the first tank should have about 70% of the volume of the second. The minimum is actually an extremely shallow one, the total volume only being 3% less (for 99% reaction) than if both tanks had the same volume. Such a saving in total volume would not compensate for the extra cost of installing and maintaining two tanks of different sizes [**3, 4**].

We have so far considered examples where the reaction rate can be expressed as a simple function of one concentration. The algebra involved can become tedious if a long chain of C.S.T.R.s is envisaged. In such a case, simplifications can be obtained by various methods – both analytical and graphical [**5–9**]. We shall only consider here a simple graphical

Fig. 14. Ratio of volume of two C.S.T.R.s in series to volume of single C.S.T.R. versus conversion.

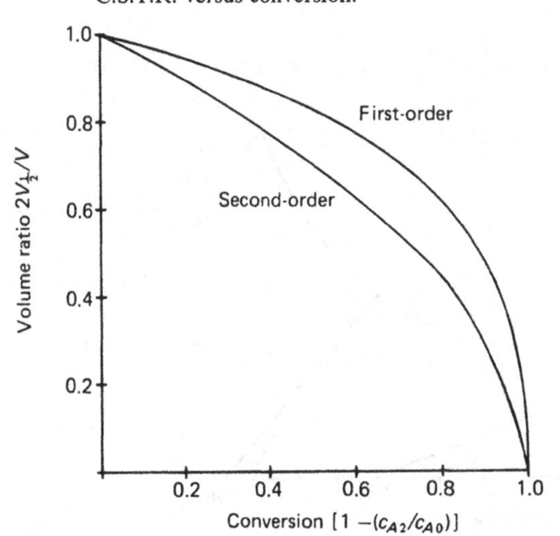

method, applicable if the rate is known as a graph against one reagent concentration. Fig. 15 shows such a graph (which may however not 'fit' any simple kinetic order expression).

Equation (2.1) tells us that for a steady state C.S.T.R. in which there is no change of density

$$vc_{A0} = vc_{A1} + rV \tag{4.12}$$

or

$$c_{A1} = c_{A0} - \frac{V}{v}r. \tag{4.13}$$

Fig. 15 shows that if we draw a line passing through the point $(c_{A0}, 0)$ with slope $-v/V$, it will intersect the rate curve at c_{A1}.

It is easy to see that a similar process can be used to find c_{A2}, if a second tank is placed in series. Although v will be the same, it is not necessary that the volume of the second tank should equal V. If the problem is to get from c_{A0} to c_{A2} by using two tanks, then c_{A1} will also be a matter of choice. Optimization by trial and error is a simple matter.

4.4 Autocatalytic reactions

Although it is true in the great majority of cases that a C.S.T.R. will have a larger volume than the comparable tubular reactor, this is not invariably so. A reaction of zero order would require a C.S.T.R. of the same volume as that of the comparable tubular reactor. Apart from this rather academic example, there is another important class of reactions in which a C.S.T.R. can provide advantages in reducing the volume

Fig. 15. Graphical method for C.S.T.R. design.

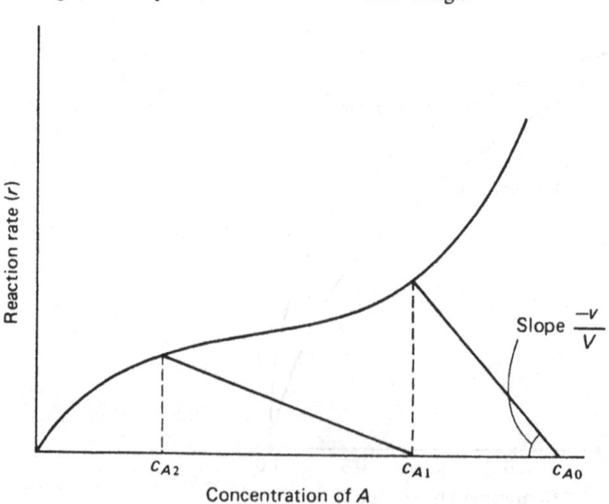

needed *below* that required in a tubular reactor. This class is that of autocatalysed reactions, in which the products of reaction increase the rate of reaction. As an example, consider Fig. 16. This shows how the reaction rate in a liquid containing reagent A, initially at a concentration c_{A0}, may increase as the concentration of A falls. This is because of the increase in the concentration of reaction products, which autocatalyse the reaction. Eventually the rate falls again as c_A diminishes to zero.

Strictly speaking, if the reaction is autocatalytic, the rate should be zero if there is present no product whatsoever, i.e. the point P, representing the starting rate, should be on the c_A axis. In other words, the reaction should never get started. However it frequently occurs that some other, uncatalysed, reaction path is available and that this will proceed until enough products are formed for its further effect to be negligible in comparison with the autocatalytic path. In practice, small amounts of reaction product left in the reactor from previous runs may suffice to get the reaction going again and in any case reaction product can always be added to the feed.

The advantage of a C.S.T.R. in this situation of autocatalysis is that it continually mixes the incoming unreacted feed with product. It is thus possible to operate a C.S.T.R. *all the time* at the point of highest reaction rate, i.e. at point Q, and thereby to achieve a lower volume than required for a batch or tubular reactor, which would operate over a range of varying reaction rates on either side of the optimum. However this would entail an outlet concentration from the C.S.T.R. equal to c_{A1}. If it is desired to reduce c_A to a low value, the best design solution (in the sense

Fig. 16. Rate curve for an autocatalytic reaction.

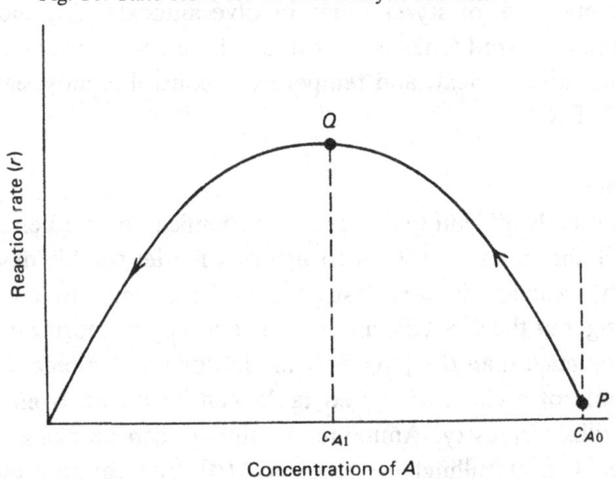

of minimum total volume) would be to use a C.S.T.R. to reduce c_{A0} to c_{A1}, and pass the effluent from the C.S.T.R. through a tubular reactor to reduce c_A to the required low limit in the final product.

Example 4.8

The reaction $A \rightarrow B$ is carried out in dilute solution. This reaction is autocatalytic, the reaction rate being given by $k[A][B]$. A solution containing 1 kmol m^{-3} of A is to be 99% decomposed in a reactor system. 10 m^3 h^{-1} of the solution is to be treated, and at the reaction temperature $k = 4.2 \times 10^{-4}$ m^3 kmol^{-1} s^{-1}. Calculate the minimum total reactor volume.

[*Answer.* 43.6 m^3.]

Example 4.9

The reaction of Example 4.8 is to be similarly carried out, but this time a tubular reactor with recycle of part of the product stream will be employed. What recycle ratio (volume of recycle/volume of feed) will lead to minimum volume of the reactor? What is this minimum volume?

[*Answers.* 0.19, 49.3 m^3.]

There are many examples of autocatalytic behaviour in biochemistry. Other chemical reactions, sometimes involving chain mechanisms, exhibit this behaviour. A product of reaction (either an intermediate or the final product) catalyses the main, or propagation, reaction. It may be advantageous to add other catalysts, such as transition metal ions. The oxidation of many organic compounds to produce such products as phenol, acetaldehyde, acetic acid, or styrene may involve autocatalysis, and in such cases the use of stirred tanks is indicated. These reactions can also evolve large quantities of heat, and temperature control is most easily effected in a C.S.T.R.

4.5 Summary

It is not usually difficult to ensure that a chemical reactor behaves as a C.S.T.R., if the designer wishes to use this model for his design equations. If the reaction is very fast, it may be difficult to ensure 'complete' mixing, but the C.S.T.R. model can generally be more closely approached in practice than the plug-flow model for tubular reactors.

A final example of a chain of stirred tanks can be drawn from the 'traditional' fertilizer industry. Ammonium sulphate can be made, for example by the I.C.I. at Billingham, England [**10**], by treating ground anhydrite ($CaSO_4$) with 'carbonated liquor', ammonium carbonate solution. Ammonium sulphate was a valuable cheap fertilizer. The reaction is, in simple terms,

$$(NH_4)_2CO_3 + CaSO_4 \rightarrow CaCO_3 + (NH_4)_2SO_4.$$

The calcium carbonate is less soluble than the calcium sulphate and hence the sulphate particles are converted to the carbonate. The solid is removed by filtration, and evaporation gives ammonium sulphate.

With the advent of newer routes to ammonia, from natural gas, and with the increase in energy prices, this process is no longer economic, but when it was in operation it provided an interesting example of the kinetics of solid dissolution and crystallization. To provide adequate contact, the slurry of solid particles in the liquor had to be stirred. To produce nearly complete conversion of the anhydrite to calcium carbonate in a single tank would have required an immense volume. A chain of six tanks in series, the process stream flowing from one to the next by gravity (the liquid level was about a foot lower in each successive tank) produced virtually complete conversion in an acceptable volume. The tanks were still a formidable size, but so was the output of the plant – of the order of a million tonnes per year.

A somewhat similar example was more recently the site of a major disaster. In 1974 there was a devastating explosion at Flixborough, in England, which killed twenty-eight people and caused many other injuries and much damage. The explosion arose from a massive leak of vapour from a plant oxidizing cyclohexane. This plant consisted of six C.S.T.R.s in series, through which the cyclohexane passed by gravity, each C.S.T.R. being set 35 cm lower than its predecessor in the train. The cyclohexane emerging from the plant was converted to cyclohexanol + cyclohexanone only to some 6%. This low conversion was for kinetic and selectivity reasons – of a type which will be discussed in Chapter 6. The liquid stream contained catalyst, and air was sparged through each reactor, thus the gas stream was 'cross-flow', the liquid stream being 'through-flow'. A leak from C.S.T.R. number 5 led to its removal, a connecting pipe between reactors 4 and 6 being constructed while reactor 5 was examined and repaired. The Court of Inquiry [**11**] held that it was the faulty design of this connecting pipe which led to the disaster.

In recent years much increased attention has been paid to the 'hazard and operability' study of chemical plants. Processes such as that at Flixborough bring together all aspects of the chemical reaction engineer's expertise. At first sight, such a plant seems a simple chain of six stirred tanks. However, the chemistry is complex, as are the mixing and mass transfer between gas and liquid phases. As well as these interesting technological problems, a proper consideration of the safety of the plant requires the combined knowledge of those expert in materials, mechanical engineering, and hazard analysis – as well as the reaction engineer. Care at the design stage should be matched by care in operation, and modifications to the plant should be considered as carefully as the original design.

Example 4.10

A reaction $A + B \rightarrow$ Product takes place in solution and is second-order irreversible. The concentrations of A and B are equal, and there is no change in density. The required degree of conversion is 99%, and laboratory batch experiments show this can be attained in 10 minutes. What would be the necessary ratios of V/v (average holding time) for (a) a single C.S.T.R., (b) two equal-volume C.S.T.R.s in series.

[*Answers.* 1000 minutes, 79 minutes.]

Example 4.11

The hydrolysis of an ester in an aqueous alkali solution is first order with respect to each of the two reagents. In a particular instance the process is carried out with a large excess of alkali in a cascade of five vessels each holding an equal volume of solution and held at the same temperature. The ratio of concentrations, e_1/e_2, of the ester in the first and second tanks is found to be the following function of the speed of stirring.

Speed (rev min^{-1})	10	20	50	100	200	300
e_1/e_2	1.44	1.49	1.54	1.57	1.58	1.58

What is the probable cause of this effect of stirring speed?

Estimate the degree of conversion obtainable in the whole cascade at the highest stirring speed. If the same process were carried out in a tubular reactor having a volume equal to the total of the five vessels and operating under the same conditions of flow rate and temperature, make an estimate of the maximum percentage hydrolysis of the ester which could be attained. Why might the actually achieved conversion be somewhat less?

[*Answers.* 89.9%; 94.5%.]

Example 4.12

A stirred tank reactor volume 10 m^3 is initially empty and is fed with two separate streams, which are well mixed in the tank. One stream is supplied at 5 m^3 h^{-1} and contains an ester at concentration 1.0×10^{-3} kmol m^{-3}; the other stream is supplied at 10 m^3 h^{-1} and contains caustic soda at concentration 0.5 kmol m^{-3}. The irreversible hydrolysis reaction occurs throughout the filling process and has a second-order rate constant of 0.1 m^3 $kmol^{-1}$ s^{-1}. The reaction may be treated as pseudo-first order in ester.

Show that during the filling period the concentration of ester, c, at a time, t, in the tank may be written

$$c = \frac{\alpha}{\beta t}(1 - e^{-\beta t}),$$

where α and β are appropriate constants.

Calculate;

(a) the concentration of ester in the solution which first leaves the tank when it is full;

(b) the concentration of ester leaving the tank when steady state conditions have been established.

[*Answers.* (a) 4.167×10^{-6}; (b) 4.115×10^{-6} kmol m^{-3}.]

Example 4.13

A chain of six equal C.S.T.R.s in series is used to carry out the isothermal first-order decomposition of a reagent in solution. Under normal operating conditions the reagent in the stream leaving the plant is at a concentration equal to 5% of that entering the plant.

Due to a malfunction of the first reactor, it is removed from the chain, the feed entering the second C.S.T.R.

(a) What will now be the concentration of reagent in the product, if the feed rate is the same?

(b) What changed feed rate would give the same product concentration as originally?

[*Answers.* (a) 8.24%; (b) reduce rate to 0.79 of former.]

Symbols

- c Concentration, kmol m^{-3}.
- G Mass flow rate, kmol s^{-1}.
- k Velocity constant of reaction, units depend on kinetics.
- P Pressure, atm.
- r Reaction rate, kmol m^{-3} s^{-1}.
- R Gas constant, atm m^3 $kmol^{-1}$ K^{-1}.
- t Time, s.
- T Absolute temperature, K.
- v Flow rate through reactor, m^3 s^{-1}.
- V Reactor volume, m^3.

References

1. Bryant, J. *Advances in Biochemical Engineering*, Vol. 5 (Springer-Verlag, 1977), p. 101.
2. Dent, F. J., *The New Scientist*, 1963, **17**, 184.
3. Wood, R. K. and Stevens, W. F., *Chem. Engng Sci.*, 1964, **19**, 426.
4. Szépe, S. and Levenspiel, O., *Ind. Eng. Chem., Process Design and Dev.*, 1964, **4**, 214.
5. Eldridge, J. W. and Piret, E. L., *Chem. Engng Progr.*, 1950, **46**, 290.
6. Jones, R. W., *Chem. Engng Progr.*, 1951, **47**, 46.
7. Wallas, S. M., *Reaction Kinetics for Chemical Engineers* (McGraw-Hill, 1959), p. 83.
8. Provinteev, I. V., *Zh. Prikl. Khim., Leningr.* 1951, **24**, 807.
9. Bilous, O. and Piret, E. L. *A.I.Ch.E. Journal*, 1955, **1**, 480.
10. Manning, J., *Chemistry and Industry*, ed. D. G. Jones (Clarendon Press, Oxford, 1967), p. 56.
11. *The Flixborough Disaster.* Report of the Court of Inquiry (H.M.S.O. London, 1975).

5

Residence-time distributions, mixing and dispersion

5.1 The residence-time distribution as a factor in reactor performance

We have seen in the preceding chapters that the volume of a C.S.T.R. to convert a given feed to a given product may be very much larger than that of the comparable tubular reactor. An important reason for this is the 'bypassing' effect, whereby a part of the feed spends a shorter time than the average in the C.S.T.R., and this is usually not compensated, in terms of product made, by the part which stays in for longer than the average. A spread of 'residence-times' in any reactor is in most cases disadvantageous, in that the reactor has to be larger than the comparable plug-flow reactor, in which there is, by definition, no spread of residence-times.

To put the problem another way, the performance of a reactor of known volume, V, treating a feed supplied at rate v (volume per unit time) is not solely determined by V, v, the feed composition, and the velocity constant. Two simple limiting cases – the plug-flow reactor and the C.S.T.R. – have been discussed, and design calculations are straightforward in these cases. In general, however, a reactor will not have the characteristics of either of these two limiting cases. It is the purpose of this chapter to show how estimates can be made of the performance of any reactor, by making use of the techniques of residence-time measurements.

There are two uses to which such procedures can be put. First, they can be used to give reasonably accurate predictions of the performance of reactors which do not approximate to either of the two limiting cases already discussed. Secondly, they can be used to determine rather more precisely the performance of reactors which, though designed to approximate to either plug-flow or C.S.T.R. behaviour, in practice deviate to some extent from such behaviour.

5.2 Residence-time functions, and relations between them

We shall assume that we have an item of equipment through which matter is passing at a steady rate. In practice, small oscillations of flow rate about a set value often occur in chemical plants, but we shall not consider here such variations. Although the overall flow rate may present no detectable variation, it is, of course, not true that all elements of matter passing through the system stay in the system for the same length of time. There is, in any practical case, a spread of residence-times.

To describe this spread of residence-times we shall define certain functions. We shall use the notation of Danckwerts [**1**], which has been used by several other authors. It is convenient to confine the discussion initially to *non-reacting* systems. The extension of the results to reacting systems will be made later.

(a) The residence-time distribution function, $E(t)$

This is defined by saying that the fraction of material in the *outlet* stream which *has been* in the system for times between t and $t+dt$ is equal to $E\,dt$. E is a function of t, $E(t)$, and clearly

$$\int_0^\infty E(t)\,dt = 1.$$

It is equally possible to define the residence-time distribution function in terms of the inlet stream. That is to say, $E(t)\,dt$ is the fraction of the *inlet* stream which *will spend* a time between t and $t+dt$ in the system. For a system in which the flow rate is constant it can be shown that $E(t)$ is the same whether defined in terms of the inlet or the outlet stream [**2**], though this is not so if the flow rates in and out are not steady. For steady flows we shall choose whichever definition of $E(t)$ is the more convenient.

(b) The cumulative residence-time distribution function, $F(t)$

The fraction of material in the outlet stream which has been in the system for times less than t is equal to F. F is a function of t, $F(t)$; clearly $F(0)=0$ and $F(\infty)=1$. It can also be seen that

$$F(t) = \int_0^t E(t')\,dt' \quad \text{or} \quad E(t) = dF(t)/dt. \tag{5.1}$$

(c) The internal-age distribution function, $I(t)$

The fraction of the material *within* the system (note *not* in the outlet stream) which has been in there for times between t and $t+dt$ is equal to $I(t)\,dt$. For a perfect mixer, and for this case only, $I(t)=E(t)$, because the outlet stream is representative of the uniform contents of the system.

(d) The mean residence-time, $\bar{t}$

The average time spent by material flowing at rate v through a volume V can be shown to be equal to V/v; i.e. the mean residence-time, $\bar{t}$, equals V/v. This statement requires two conditions. First, there must be no change of density of the flowing stream as it passes through the system (see Chapter 3). Secondly, there must be no back-mixing of material in the system upstream past the feed point, or of the product stream back into the system. Where this may occur, allowances can be made [**3, 4**].

From the definition of what is meant by the residence-time distribution, $E(t)$, it is clear that the mean residence-time is also given by

$$\bar{t} = \int_0^\infty tE(t)\,dt. \tag{5.2}$$

The following relationships, which are 'normalization equations', will now be listed. The reader should justify them for himself (a mass balance at time t on fluid which entered at time zero will prove useful).

$$\text{(i)}\quad \int_0^\infty E(t)\,dt = \int_0^\infty I(t)\,dt = 1, \tag{5.3}$$

$$\text{(ii)}\quad I(t) = \frac{v}{V}[1 - F(t)], \tag{5.4}$$

$$\text{(iii)}\quad \int_0^{V/v} F(t)\,dt = \int_{V/v}^\infty [1 - F(t)]\,dt. \tag{5.5}$$

It is sometimes convenient to work in terms of a 'reduced' or 'dimensionless' time, θ, where $\theta = vt/V$. This change of time scale has the following effects:

(1) The mean residence-time, $\bar{\theta}, = 1$.

(2) F remains unchanged, i.e. $F(\theta) = F(t)$, in which the values of θ and t are related to each other by $\theta = vt/V$.

(3) E and I become dimensionless and are V/v times as large as they were on the t scale. Thus $E(\theta) = (V/v)E(t)$ where, again, θ and t refer to the same time on the two scales of time measurement.

As an example, equation (5.4) becomes

$$I(\theta) = 1 - F(\theta). \tag{5.6}$$

Care should thus be exercised over the time scale in using results quoted in the literature.

Example 5.1

Draw sketches of E, F and I for fluid flowing in plug flow at rate v through a volume V. Use both the t scale and the θ scale.

5.3 Residence-time in a chain of stirred tanks

Let us consider a chain of n stirred tanks, each of volume V_0 m^3, through which fluid is passing steadily at a rate v m^3 s^{-1}. Suppose that, at time $t=0$, q_0 moles in the first tank are 'tagged'; these molecules become identifiable, but otherwise have the properties of the rest of the fluid. They do not react, and their concentration can be measured anywhere in the system. (If the q_0 moles were very rapidly added 'at' $t=0$, this would start a 'tracer' experiment, of which we shall have more to say later in this chapter.) Our present problem is to calculate the quantities q_1, q_2, etc., of the tagged molecules present in the various tanks at any later time t.

Assuming perfect mixing, the concentration of tracer in the first tank at time t is q_1/V_0 and the rate of outflow is vq_1/V_0. There is no inflow term and no reaction term. The mass balance equation for this tank is therefore

$$-\frac{dq_1}{dt}=\frac{vq_1}{V_0}. \tag{5.7}$$

Hence

$$q_1=q_0\,e^{-vt/V_0} \tag{5.8}$$

Similarly, for the second tank

$$-\frac{dq_2}{dt}=\frac{vq_2}{V_0}-\frac{vq_1}{V_0}, \tag{5.9}$$

where the first and second terms on the r.h.s. represent the rates of outflow and inflow respectively. Substituting (5.8) in (5.9) to eliminate q_1, we obtain after re-arranging

$$\frac{dq_2}{dt}+\frac{vq_2}{V_0}=\frac{v}{V_0}q_0\,e^{-vt/V_0}. \tag{5.10}$$

The integrating factor for this equation is e^{vt/V_0} and the solution is

$$q_2=\frac{vt}{V_0}q_0\,e^{-vt/V_0}. \tag{5.11}$$

Proceeding in the same way for the third tank we obtain

$$q_3=\left(\frac{vt}{V_0}\right)^2 q_0\frac{e^{-vt/V_0}}{2!}. \tag{5.12}$$

Consider now the fluid leaving the first tank between times t and $t+dt$. The quantity of our tracer material in this fluid is vq_1dt/V_0, since the concentration of tracer is q_1/V_0. Thus the fraction of the original tracer leaving the first tank between t and $t+dt$ is given by

$$\frac{vq_1dt/V_0}{q_0}=E(t)\,dt. \tag{5.13}$$

Substituting for q_1 from equation (5.8), we obtain

$$E(t)=\frac{v}{V_0}e^{-vt/V_0}, \tag{5.14}$$

which is the residence-time distribution function for a single well-stirred tank.

***Example* 5.2**

Draw sketches of E, F and I for fluid flowing at rate v through a well-stirred tank of volume V_0. Compare the results on the t and θ scales with those from Example 5.1.

***Example* 5.3**

Derive the residence-time distribution function $E_3(t)$ for three stirred tanks in series, using equation (5.12).

If we now return to equations (5.11) and (5.12) it is possible to extend the treatment, so that for the ith tank

$$q_i=\left(\frac{vt}{V_0}\right)^{i-1} q_0 \frac{e^{-vt/V_0}}{(i-1)!}. \tag{5.15}$$

The solutions to equations (5.10), (5.11) and (5.12) have been plotted in Fig. 17. This shows how the tracer substance moves progressively through a series of three tanks and how the tank having the highest concentration is initially the first, later the second and finally the third.

Fig. 17. Progressive movement of a tracer substance through a chain of stirred tanks.

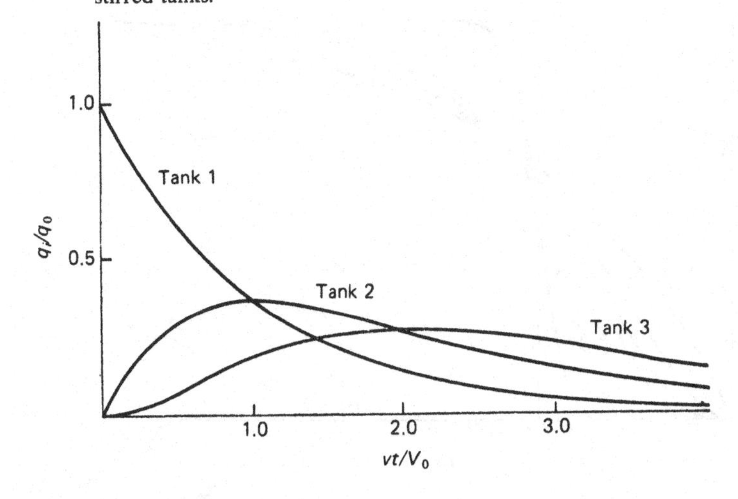

The quantity of tracer which has 'escaped' from a set of n tanks during the time t since its addition is

$$q_0-(q_1+q_2+\cdots+q_n).$$

The fraction of the original tracer which has escaped during this time is equal to F, as consideration of § 5.2a and 5.2b will show. We therefore have

$$\begin{aligned}F&=\frac{q_0-(q_1+q_2+\cdots+q_n)}{q_0}\\&=1-e^{-vt/V_0}\left[1+\frac{vt}{V_0}+\left(\frac{vt}{V_0}\right)^2\frac{1}{2!}+\cdots+\left(\frac{vt}{V_0}\right)^{n-1}\frac{1}{(n-1)!}\right].\end{aligned} \tag{5.16}$$

The mean residence-time for the whole chain of n tanks is $\bar{t}=nV_0/v$. If we substitute this into equation (5.16) we obtain the more convenient form

$$F=1-e^{-nt/\bar{t}}\left[1+\frac{nt}{\bar{t}}+\left(\frac{nt}{\bar{t}}\right)^1\frac{1}{2!}+\cdots+\left(\frac{nt}{\bar{t}}\right)^{n-1}\frac{1}{(n-1)!}\right]. \tag{5.17}$$

This equation gives the fraction F of the tracer which has escaped into the outflow during the time t, as a function of n, the number of tanks, and of the ratio $t/\bar{t}$ (which is the same as the quantity θ, mentioned in § 5.2d). This function is shown in Fig. 18, and in addition some values of the percentage bypassing (i.e. $100F$) are given in Table 1.

Fig. 18. The fraction of tracer substance which, at time t, has passed out of a system consisting of n equal stirred tanks in series.

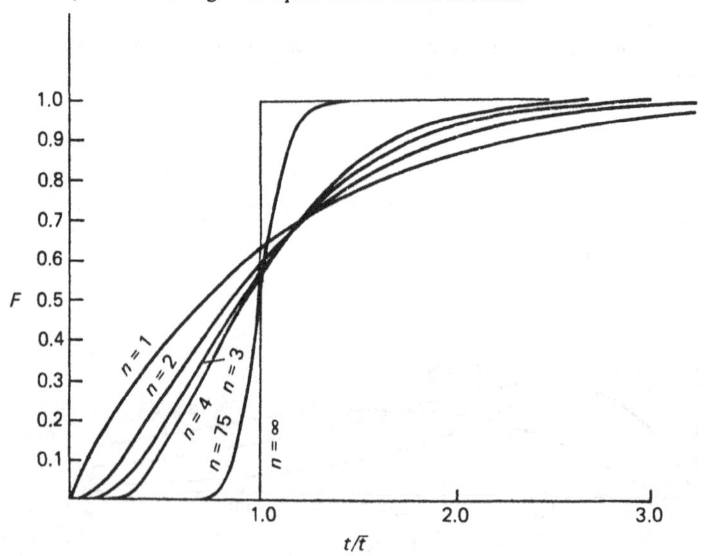

Table 1

	$n=1$	$n=2$	$n=3$
$t/\bar{t}=0.1$	9.5	1.8	0.4
$t/\bar{t}=1.0$	63.2	59.4	57.7

Considering the first row of the table it will be seen how the effect of increasing the number of tanks is greatly to reduce the percentage of the tracer which has found its way into the outlet from the system in a time short compared to the mean residence-time. Considering the second row it will be seen that *more than* 50% of the tracer escapes in a time equal to the mean residence-time. Thus, for a single tank, 63% passes out between $t=0$ and $t=\bar{t}$ and the remaining 37% passes out between $t=\bar{t}$ and $t=\infty$. In regard to the definition of the mean residence-time, a smaller percentage of molecules having residence times from $\bar{t}$ up to infinity exactly compensates a larger percentage having residence times between zero and $\bar{t}$.

Apart from the supposition that the liquid flow is steady and is constant from tank to tank, the only other assumption in the above theory is perfect mixing. Therefore an experimental measurement of F and its comparison with the curves of Fig. 18 can be a valuable means of obtaining an indication of the extent to which there is an approach to perfect mixing in the system in question.

5.4 Residence-time distributions for composite systems

We have seen in the previous section how the residence-time distribution for two stirred tanks in series can be derived (see equation (5.11)). The present problem is as follows: given a unit for which the residence-time distribution is $E_1(t)$, and another unit for which we have $E_2(t)$, what will be the residence-time distribution function, $E_{1+2}(t)$, for the two units together in sequence? Here neither of the units, of course, need be a stirred tank.

The derivation will not be given in detail, but will be left as an exercise for the reader. Consider a tracer element entering the first unit at $t=0$. It is 'split up' into sub-elements which leave the first unit at various times thereafter. Each of these sub-units can be regarded as a separate 'tracer element' for the second unit.

When analysing the effluent from unit 2 we will not be able to distinguish between tracer which spent times t_1 and t_2 in units 1 and 2, and tracer which spent times t_1' and t_2' in units 1 and 2, where $t_1+t_2=t_1'+t_2'=t$. Nor is such distinction relevant to the definition of $E_{1+2}(t)\,dt$, which will include contributions from all possible combinations of $t_1'+t_2'$.

Hence it can be shown that

$$E_{1+2}(t)\,dt=\int_0^t E_1(t_1')E_2(t-t_1')\,dt_1'\,dt. \tag{5.18}$$

Example 5.4

Use equation (5.18) to show that $E(t)$ for two well-stirred tanks in series, each of volume V_0, through which the flow rate is v, is given by

$$\left(\frac{v}{V_0}\right)^2 t\,e^{-vt/V_0}.$$

Compare this with the derivation of equation (5.11). Show also that the mean residence time is $2V/v$, and that $E(\theta)=4\theta\,e^{-2\theta}$.

Example 5.5

Show by induction that $E(t)$ for a chain of n well-stirred tanks in series, each of volume V_0, through which the flow rate is v, is given by

$$\left(\frac{v}{V_0}\right)^n \frac{t^{n-1}}{(n-1)!}\,e^{-vt/V_0}.$$

Convert this to $E(\theta)$, where θ is based on the mean residence-time in the whole chain of n tanks.

5.5 The determination of residence-time distributions

In previous sections we have described how the residence-time distribution function can be derived for some model systems. In practice the reverse procedure is of greater interest. The residence-time distribution is determined experimentally, and information about the system is derived from these results. For example, residence-time measurements on a tubular reactor can be used to discover how closely, or otherwise, the plug-flow assumptions are in practice obeyed.

The methods of determining residence-time distribution functions in flow systems will now be described. Our comments will be confined to three basic methods – 'pulse signal', 'step-change', and 'periodic input'.

(*a*) *Pulse signal*

In this method a quantity q_0 of tracer is added to the ingoing stream during (and only during) a period of time which is very short compared with $\bar{t}$. The concentration of this tracer material is measured in the outlet stream as a function of time.

Tracers which have been used for this purpose are legion. One can choose any tracer most convenient for the given system, provided it satisfies certain conditions. It should not affect the flow; it should be conserved (i.e. a mass balance on it must be possible); it should be able to be injected in a short time; and it should be conveniently analysed. Its molecular diffusivity should be low (see later) and it should not be absorbed by, or react with, the surface of the 'reactor' (e.g. heavy water is not a satisfactory tracer in biological systems because it undergoes exchange processes with living tissues).

It is also necessary that the tracer be uniformly distributed in the ingoing fluid, and that the analysis at the exit should give a proper average concentration in the outgoing fluid. Where there is a variation of fluid velocity across a vessel cross-section care must be taken to ensure that the measurements give proper weight to fluid flowing with different velocities [**5**], [**6**].

We have seen in the previous section how the tracer concentration in the outlet stream is related to the E function, for a well-stirred tank. In the general case, let us consider the tracer leaving a system between t and $t+dt$. From the definition of E, see § 5.2a, this quantity of tracer is $q_0 E(t)\,dt$. From the definition of $c(t)$, the concentration of tracer in the outlet, this quantity of tracer is $c(t)v\,dt$. Therefore

$$E(t)=\frac{v}{q_0}c(t). \tag{5.19}$$

Thus apart from the scale factor v/q_0, the graph of $c(t)$ is the same as that of $E(t)$. Note that the units of $v\times c$ must be consistent with those of q_0. A typical pulse signal (or 'slug injection') curve might be as shown in Fig. 19. For this system it is seen that the distribution is fairly sharply peaked about the mean residence-time. For a 'plug-flow' reactor the peak would be extremely sharp; in the limit it would approach a Dirac delta function at the mean residence-time. The Dirac delta function, $\delta(t-t_0)$, has the property of being zero for all times except $t=t_0$ and also the property that $\int\delta(t-t_0)\,dt=1$ if the range of integration includes $t=t_0$, being zero if it does not.

***Example* 5.6**

Write down $E(t)$ for a plug-flow system in terms of the Dirac delta function. Use equation (5.18) to derive the residence-time distribution function for a plug-flow system of volume V_1 followed by a C.S.T.R. of volume V_2. Show that the result is the same as for the C.S.T.R followed by the plug-flow system. Under what circumstances would you expect the residence-time distribution function for a general system to remain unchanged if all the flows are reversed?

Example 5.7

In a tracer experiment 2 g of an inert tracer are quickly added to and dissolved in the feed stream of a reactor. The concentrations, in g m^{-3} of the tracer in the fluid leaving the reactor at various times after the moment of addition, are found to be as follows:

t (min)	0.1	0.2	1.0	2.0	5.0	10.0	20.0	30.0
concentration (g m^{-3})	1.960	1.930	1.642	1.344	0.736	0.268	0.034	0.004

The volume of fluid in the reactor is 1 m^3 and the volume rate of flow is constant at 0.2 m^3 min^{-1}.

Show, by doing a material balance, that the quantity of tracer left in the system is negligible after 30 minutes. Use the exit concentrations to estimate the mean residence-time. Does the performance of this reactor approximate at all closely to that of any simple model?

(*b*) *Step-change*

If the feed to a system is switched over instantaneously from one supply to another a 'step-change' residence-time experiment can be carried out. The second feed must be distinguishable from the first, but must behave within the system in a similar manner to the first. For example, both feeds should have the same density and viscosity. Any measurable property of the two feeds may be used, e.g. the calorific value of the gas coming from a plant which switches from one feed tank to another. It

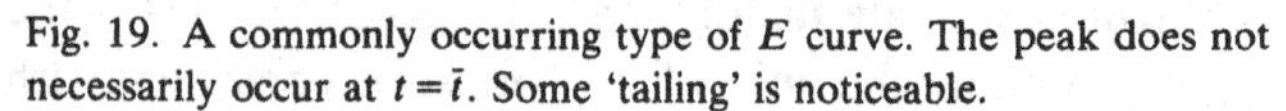

Fig. 19. A commonly occurring type of E curve. The peak does not necessarily occur at $t = \bar{t}$. Some 'tailing' is noticeable.

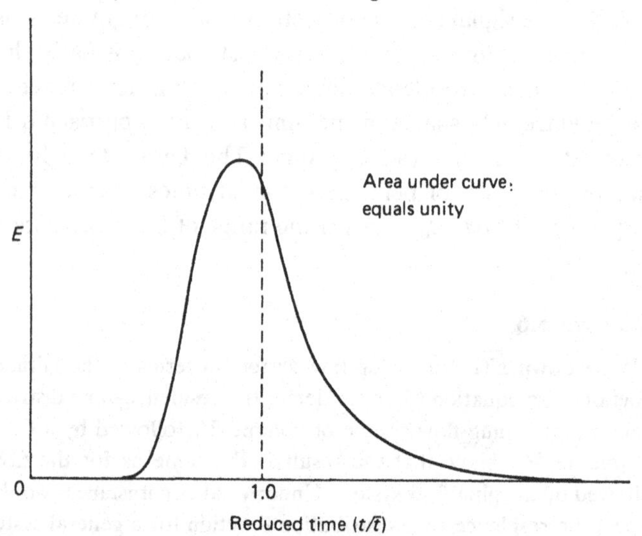

is not always necessary to add any special tracer material. However, switching from undyed to dyed feed has often been used, and such addition of a tracer material is more usual. If the tracer is expensive and/or unpleasant (e.g. radioactive tracers) a pulse experiment would be preferable.

Consider the effluent from the system at time t after the step-change to a 'marked' feed. The concentration of the tracer in the effluent is simply related to the F function of the system, since all *unmarked* material in the effluent must have been in the system for a time greater than t. Thus if P is some property which varies linearly with the volume fraction of feed 1 in a mixture with feed 2, then it is clear that

$$1-F(t)=\frac{P_2-P(t)}{P_2-P_1} \quad \text{or} \quad F(t)=\frac{P(t)-P_1}{P_2-P_1}.$$

Thus a graph of $P(t)$ against time would show a change from P_1 (before the step-change) to P_2 (a long time afterwards), and this graph would be the F curve for the system, apart from the shift of origin and change of scale.

Fig. 20 shows a typical F curve, such as might be derived from a step-change experiment. Such an F curve can, of course, be calculated from an E curve obtained by a pulse experiment, or vice versa, see equation (5.1).

The mean residence-time can be obtained by finding that time at which the two shaded areas in Fig. 19 are equal (see equation (5.5)). It is clear that in this case the spread of residence-times is not great. For a plug-flow reactor the step would be extremely sharp and in the limit it would

Fig. 20. The F curve corresponding to Fig. 19.

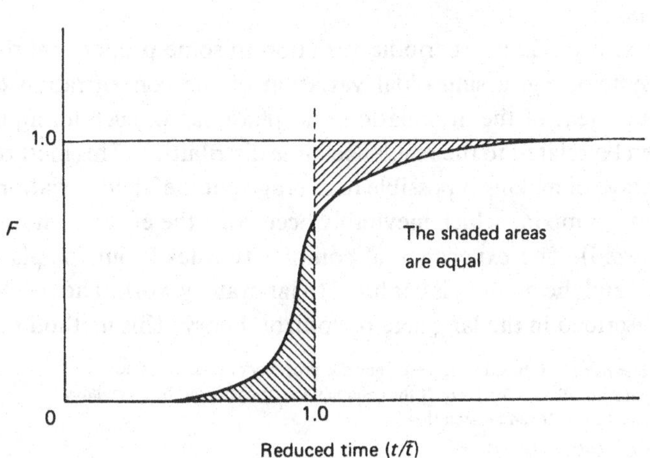

approach a Heaviside unit function* at the mean residence-time. On the other hand, the spread of residence-times may be large, as Example 5.2 shows clearly.

***Example* 5.8**

The gaseous effluent from a packed-bed reactor is continuously monitored by a spectrophotometer. The spectrophotometer reading (at a given wavelength) is proportional to the concentration of a non-reacting component in the gas. The feed to the reactor is changed from one supply to another, and the spectrophotometer readings at time t after the change are shown in the table below.

Time (s)	Reading	Time (s)	Reading
0	1.650	100	1.771
40	1.650	120	1.783
60	1.665	140	1.796
70	1.698	180	1.800
80	1.749	220	1.800
90	1.763		

Calculate the mean residence-time of the gas in the reactor. The reactor consists of a cylinder of 0.6 m internal diameter and 2 m long filled with randomly packed catalyst beads. The flow rate through the reactor is $0.25\ m^3\ min^{-1}$ (measured at reactor conditions). Calculate the voidage of the catalyst bed. Comment on your answer.

[*Answers.* 83 s, 0.61.]

(*c*) *Periodic input*

A third method is to apply a periodic variation to some property of the feed into a system, e.g. a sinusoidal variation of the concentration of tracer. Measurement of the attenuation and phase lag of such an input signal can then be related to the residence-time distribution. This method has the advantage of making it possible to average out the slight variations in feed rate and in mixing which inevitably occur (i.e. the effect of 'noise' can be minimized). The experimental complexity rules it out for plant investigations, and the method is confined to laboratory work. The results are usually described in the language of control theory. This method has

* The Heaviside unit function $H(t-t_0)$ has the property of equalling 0 for $t<t_0$ and equalling 1 for $t>t_0$. It thus enables a step-change to be formulated mathematically. It can be seen that
$\delta(t-t_0)=(dH(t-t_0)/dt)$.

been applied, for example, to the distribution of residence-times in a falling water film [**7**].

5.6 The measurement of residence-time functions in industry

The techniques of residence-time measurement are now widespread in industry. The results of such investigations may have only local value, and relatively few accounts are given in the literature. An interesting group of papers on residence-time studies on catalytic cracking plants should be mentioned [**9–10**]. Helium was used as tracer for the gas flow. Radioactive tracers were used for the catalyst, which flows in a cycle from reactor to regenerator and back to the reactor. Two of the catalytic crackers behaved fairly closely as three perfect mixers in cycle, but a third plant fitted better the model of two perfect mixers and a plug-flow stripper in cycle.

It is sometimes difficult to measure the last traces of tracer emerging from a plant; the E curve has a long 'tail'. This may not be serious, in that these last traces may not be large enough to prevent an acceptably accurate estimate of the mean residence-time (equation (5.2)). But sometimes the material balance on the tracer is poor, and the indications are that significant quantities of the tracer are staying in the plant for very long and imprecisely known times. The calculation of the mean residence-time can then be rendered inaccurate by this long 'tail'. In extreme cases, with a large 'dead space' in the system, the *apparent* mean residence-time may be much too small. An example known to the authors was a plant making glass, where the viscous melt flowed rapidly in a narrow hot zone through the plant in which nearly three-quarters of the melt volume was taken up by cooler and highly viscous material which was barely moving.

However, in spite of such difficulties, residence-time measurements can give much valuable information on the nature of the flow through a chemical plant. The total quantity of material inside a working plant is often difficult to determine accurately, and a calculation of the mean residence-time enables this quantity to be calculated if the flow rate through the plant is known. The spread of residence-times about the mean has often provided an explanation of the plant's behaviour. The effect of such a spread on the output from a reactor will be considered in the next section.

5.7 The effect of a spread of residence-times on reactor yield

Since in most cases the rate of an isothermal reaction falls off as conversion increases, the conversion due to material which stays in an isothermal reactor longer than the mean residence-time does not, in general, compensate for the product 'lost' by material which stays in the reactor for less than the mean residence-time. Thus a spread of residence-times will usually lead to a reactor performance lower than that for a plug-flow reactor. It has now to be asked how much lower.

For a first-order reaction the *fractional* degree of conversion of the reagent is independent of its concentration. Therefore, if two volumes of reaction fluid of different concentrations are mixed together, the total amount of product formed will be the same, after a given time, as it would have been if the two volumes had been kept separate. This is true only of first-order reactions. If the rate of reaction were *more* dependent on concentration than corresponds to first-order kinetics, the mixing of two quantities of fluid, one more dilute that the other, would *reduce* the total amount of product formed in a given time. Conversely, if the order were *less* than first, the product formed would be *increased* by such a mixing.

Thus the performance of a reactor will in general depend upon the mixing inside it of fluid elements of different concentration. It will also depend on whether such mixings occur early or late, relative to the mean residence-time. Unfortunately the residence-time distribution does not, by itself, give sufficient information for a precise calculation of the magnitude of this effect. Thus Example 5.6 shows that the same residence-time distribution can be obtained from two different flow-systems, which would not give the same performance, as reactors, for any reaction of other than first-order kinetics (see Examples 5.11 and 5.12 later).

For a first-order reaction we can regard the effluent from a reactor as a collection of elements of fluid which have spent appropriate times in the reactor without any mixing between the elements while in the reactor. The average concentration of reagent in the effluent, $\bar{c}$, would be given by

$$\bar{c} = \int_0^\infty c(t) \cdot E(t)\, dt, \tag{5.20}$$

where $E(t)$ is the residence-time distribution and $c(t)$ is the value of c as a function of time in a batch reaction.

***Example* 5.9**

Show that for a first-order chemical reaction the calculated yield from a well-stirred tank is the same whether equation (5.20) or the mass balance equation (4.3) is used.

***Example* 5.10**

The reactor of Example 5.7 is being used to carry out a first-order chemical reaction of velocity constant 1.8 min^{-1}. What percentage of the reagent is converted in the reactor?

[*Answer.* 90%.]

Consider now a second-order reaction of a single reactant, with velocity constant k_2, being carried out in a well-stirred tank. The reader should be able to show that equation (5.20) gives

$$\frac{\bar{c}}{c_0}=\beta\int_0^\infty \frac{e^{-\beta\psi}}{1+\psi}\,d\psi. \tag{5.21}$$

Here $\beta = v/k_2Vc_0$ and $\psi = k_2c_0t$. This integral can be solved graphically, or by use of tables of exponential integrals. Equation (5.21) will correctly give the average yield of a group of batch reactors whose products are suitably mixed to give the same residence-time distribution overall as one perfect mixer. However, the yield from such a continuous perfect mixer would *not* be given by equation (5.21), but by the equation

$$\frac{\bar{c}}{c_0}=1-\frac{1}{\beta}\left(\frac{\bar{c}}{c_0}\right)^2, \tag{5.22}$$

which is easily obtainable by the application of the mass balance equations.

The difference between equations (5.21) and (5.22), which both refer to the same residence-time distribution, illustrates what was said earlier – namely that the residence-time distribution is not in general sufficient to determine the precise yield of a reactor. The next section discusses further how much *can* be deduced from the residence-time distribution.

***Example* 5.11**

A process stream flows through a plug-flow reactor and then through a C.S.T.R. of the same volume (see Example 5.6). Show that the percentage change of a reagent which decomposes by a first-order reaction is the same as it would be if the feed passed first into the C.S.T.R. and then through the plug-flow reactor. There is no change of density or temperature of the process stream.

***Example* 5.12**

A reactant decomposing by a second-order mechanism is passed through the plug-flow reactor and the C.S.T.R. of Example 5.11. The effluent from the C.S.T.R. contains 1% of the original reactant. How much of the original reactant would be left in the process stream if it passed first through the C.S.T.R. and then through the plug-flow reactor? There is no change of density or temperature of the process stream. [*Answer.* 1.44%.]

5.8 Segregation and limits of reactor yield

There have been some attempts to define the limits within which the yield from a reactor must lie, given the residence-time distribution. For a first-order reaction the limits converge to the calculable yield, as in Example 5.11. For other reactions this is not so, as in Example 5.12. The output from a reactor with a given residence-time distribution will depend upon *where* in its flow through the reactor the process stream receives that mixing which the residence-time distribution indicates. This problem has been discussed in terms of 'segregation' [**11**] and 'minimum and maximum mixedness' [**12, 13**].

If the reaction is of order higher than one, and involves only one reagent, then mixing should occur as late in the process as possible (minimum mixedness). Example 5.12, case 1, is fairly close to that state of affairs; whereas case 2 is the opposite state of maximum mixedness, and of lowest yield of product. The difference may well be small, and in general difficult to estimate.

If the reaction is one between two or more different reactants, clearly these must be mixed together at some stage or the reaction will never take place. If this is thoroughly done before or at entry, and if the reagents are in stoichiometric proportions, then the problem is not much different from that already discussed in this section. However if the mixing of the feed is not thorough or if non-stoichiometric quantities are used, the effects of mixing in the reactor are more difficult to compute.

The degree of mixing on the molecular scale may well be different from that on a somewhat larger scale. Turbulent mixing produces eddies in which sheets of fluid are intermixed and stretched. If these sheets become thin enough molecular diffusion will rapidly even out differences. Since gaseous systems have much higher diffusion coefficients than liquids, effects of incomplete mixing at the molecular level are more likely to be noticed with liquid systems than with gaseous systems.

However, if the reactants are adequately mixed at entry to the reactor, and the residence-time distribution is known, equation (5.20) will give an estimate of the reactor performance which, though usually too high (since it refers to minimum mixedness or complete segregation), will probably be acceptably close to that found.

The above conclusion applies when there is only one reaction going on in the reactor, but there are many situations of importance where this is not the case. The desired reaction, e.g. a mono-substitution, may be followed by further, consecutive substitutions; or the desired reaction may proceed in parallel with an undesired decomposition. The latter is frequently the case with organic oxidation reactions, and raises particularly severe problems in the design, and scale-up, of gas–liquid reactors. Even in single-phase reactors the selectivity obtained may depend markedly on the mixing at the molecular level, or 'micromixing', as has been shown theoretically and experimentally [**14**], [**15**].

***Example* 5.13**

For the residence-time distribution and kinetics of Example 5.12 calculate the fraction of reagent left undecomposed for the case of minimum mixedness.

[*Answer.* 0.96%, which is very close to the 1% of Example 5.12, case 1.]

5.9 Residence-times and dispersion in flow through pipes

The flow of fluids through pipes is a commonplace occurrence in chemical plants, oil refineries, supply lines, etc. It is a fact that the plug-flow assumption is not in general correct; indeed it may be so *in*correct as to lead to useless results. However, it is often the case that useful results can be obtained by employing minor corrections to plug-flow.

For this purpose we shall consider what happens to a non-reacting tracer material originally confined within a narrow disc perpendicular to the axis of a cylindrical pipe. This tracer material is dispersed as the fluid flows down the pipe, and as it passes an observer downstream (or issues from the end of the pipe) its concentration can be measured and used to determine the residence-time distribution (i.e. as in a pulse signal experiment). We shall also discuss the effects of such a distribution on any chemical reaction which may be going on in the pipe.

Consider first a non-diffusing tracer 'injected' into a (Newtonian) fluid in laminar flow.

(*a*) *Non-diffusing tracer, laminar flow*

If the parabolic velocity profile is established ('entry length' phenomena will not here be considered) the linear velocity u at the radial position r is given by the well-known equation

$$u = 2\bar{u}(1 - r^2/a^2), \tag{5.23}$$

in which it will be assumed that the average velocity, $\bar{u}$, and a, the pipe radius, are constant.

We require to find an expression for that fraction of the exit fluid which has spent time less than t in a pipe of length l – in other words $F(t)$, as defined in equation (5.1). Consider that radius, r, at which the fluid spends time t inside the pipe. All fluid which flows between the pipe centre and r will take less time than t to pass through, and this fluid amounts in quantity to q, where

$$q = \int_0^r 2\pi u r'\,dr'. \tag{5.24}$$

Substituting from equation (5.23) we obtain

$$q = 4\pi\bar{u}\left(\frac{r^2}{2} - \frac{r^4}{4a^2}\right). \tag{5.25}$$

Since the total fluid flow through the pipe is $\pi a^2\bar{u}$, we have

$$F = \frac{q}{\pi a^2 \bar{u}} = \frac{2r^2}{a^2} - \frac{r^4}{a^4}. \tag{5.26}$$

From equation (5.23) it follows that

$$\frac{r^2}{a^2} = 1 - \frac{u}{2\bar{u}} = 1 - \frac{t_0}{t}, \tag{5.27}$$

where t_0 is the time for the centre-line fluid to pass through the pipe, which fluid moves with velocity $2\bar{u}$. Substituting from equation (5.27) in (5.26) we finally arrive at

$$F = 1 - \left(\frac{t_0}{t}\right)^2 \tag{5.28}$$

Some values of F are as follows:

t/t_0	1.5	2.0	5.0	10.0
F	0.55	0.75	0.96	0.99

In other words only 1% of the fluid has a time of passage exceeding $10t_0$, or five mean residence-times.

The performance of a tubular reactor in which there is laminar flow can be estimated by differentiating equation (5.28) to obtain $E(t)$ and substituting in equation (5.20). As mentioned before, this procedure will overestimate the conversion for reactions of order >1. It must be noted that this estimate assumed complete segregation, which in view of the laminar velocity profile seems reasonable (the fluid annuli moving with different velocities do not mix). It also assumes no diffusion of reactants or products radially (which we also assumed in our tracer model to establish equation (5.28)). It will be seen later how the situation is changed by diffusion. The calculation of $\bar{c}$ using equation (5.20) will clearly depend on the kinetic equation determining $c(t)$.

***Example* 5.14**

A first-order chemical reaction is to be carried out in a tubular reactor. There will be no change of temperature or density in the reactor and 90% of the incoming reagent is to be decomposed. How much larger than the plug-flow model would the actual reactor have to be if in the latter a parabolic laminar velocity profile were established.

$$\left(\int_1^\infty \frac{e^{-\alpha T} dT}{T^3} = 0.05 \quad \text{when} \quad \alpha = 1.60.\right)$$

[*Answer.* 1.39 times as large.]

For fairly high conversions a reactor in which the laminar velocity profile obtains has to be some 30–40% larger than would be expected on the basis of a simple plug-flow calculation (the exact figure depends on the kinetic order and the desired degree of conversion). This does not allow for the effect of diffusion, and this is much more significant with gases than with liquids. Nor have we considered temperature differences arising from reaction; these can cause convectional mixing. These factors have been examined by several authors (e.g. [**16**]), and their effect is generally to bring performance closer to the plug-flow estimate.

The effect of diffusion upon the spreading of a tracer material as it flows down a pipe was considered by Taylor [**17–19**]. This work has been refined and extended by others, but the conclusions can be simply summarized. This spreading, or 'dispersion', depends upon the flow velocity; we shall first consider what happens in laminar flow.

(*b*) *Diffusing tracer, laminar flow*

Provided the time, t, between the injection of the tracer and the observation of the tracer obeys the criterion

$$Dt/a^2 \gg 0.1, \tag{5.29}$$

then the tracer will be found to be dispersed about an origin in a frame of reference moving with the *mean* velocity of the fluid, according to the relation

$$c = \frac{M}{2\sqrt{(\pi K t)}} e^{-(x-\bar{u}t)^2/4Kt}. \tag{5.30}$$

This is an expression of the form found in *diffusion* problems. $M=$ quantity of tracer per unit area of the source plane, $(x-\bar{u}t)$ is the distance from an origin moving with the mean velocity of the fluid, and K, the apparent diffusion coefficient, is given by [**20**]

$$K = D + (a^2\bar{u}^2/48D). \tag{5.31}$$

Diffusion in the radial direction is fast enough to keep the concentration, c, at any value of x substantially independent of r. It commonly occurs, especially with liquids, that the second term on the r.h.s. of equation (5.31) is much greater than the first. We thus have the curious situation in which a higher *molecular* diffusion coefficient leads to a lower *apparent* diffusion coefficient. The faster-moving fluid in the centre of the tube moves *through* the dispersed tracer, picking it up from the slower-moving fluid on the upstream side and giving the tracer back to the slower-moving fluid on the downstream side. A high *molecular* diffusion coefficient makes this 'sideways' transfer rapid. It therefore requires a short length of the tube and the *apparent* diffusion coefficient is thus low. The tracer as a whole moves downstream at a speed equal to $\bar{u}$, spreading out as it proceeds. Fig. 21 shows this dispersion, and Fig. 22 shows some experimental measurements of K [**21**].

If the diffusion coefficient, D, is very low, then K is very large – but it should be noted that curves of the type shown in Fig. 21 will only be obtained after a long time, see equation (5.29). Put another way, the pipe will have to be very long for the dispersion of tracer to obey equation (5.30) at the exit from the pipe. For a given length of pipe, the higher the mean velocity (while still keeping in the laminar regime) the less likely are equations (5.30) and (5.31) to be obeyed, and the more likely

Fig. 21. The dispersion of a disc of tracer – see equation (5.30). x, $\bar{u}t$, $\sqrt{Kt}$ and M/c must all be in the same units.

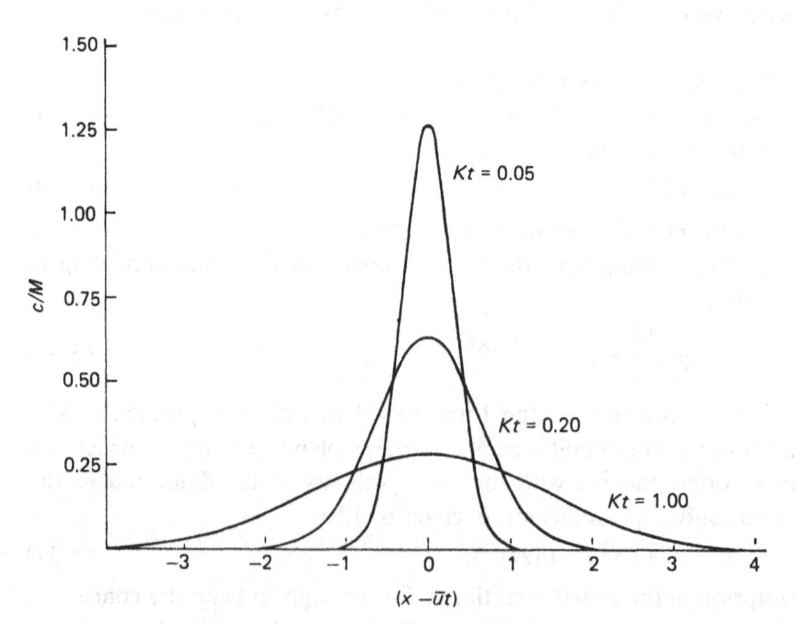

will equation (5.27) be obeyed. To see where the boundary between these two regions lies (it is not, of course, a sharp boundary) it is convenient to rewrite equation (5.29) in the form

$$\bar{u} \ll 10DL/a^2. \tag{5.32}$$

In other words, if $\bar{u}$ is much *greater* than $10DL/a^2$, equation (5.27) will be obeyed; molecular diffusion can then be neglected, and the tracer will be dispersed as a paraboloid shell such that the amount of tracer per unit length of pipe is constant, at any time, between the original source position and the nose of the paraboloid $2\bar{u}t$ downstream [**17**]. This is much more readily obtained with liquids than with gases, since the diffusion coefficients of gases are so much higher than those of liquids. It must be remembered that $\bar{u}$ must not be so high that turbulent flow results.

(*c*) *Steady turbulent flow*

In this case the eddy diffusivity is so great that the molecular diffusion coefficient is not very important by comparison. The effect of the turbulence is to flatten the velocity profile and reduce dispersion. Taylor's

Fig. 22. Plate height versus mean velocity for the dispersion of N_2/C_2H_4 mixtures in a 0.64 cm diameter straight pipe. The curves drawn are theoretical, derived from equation (5.31). Plate height $h = 2K/\bar{u}$.

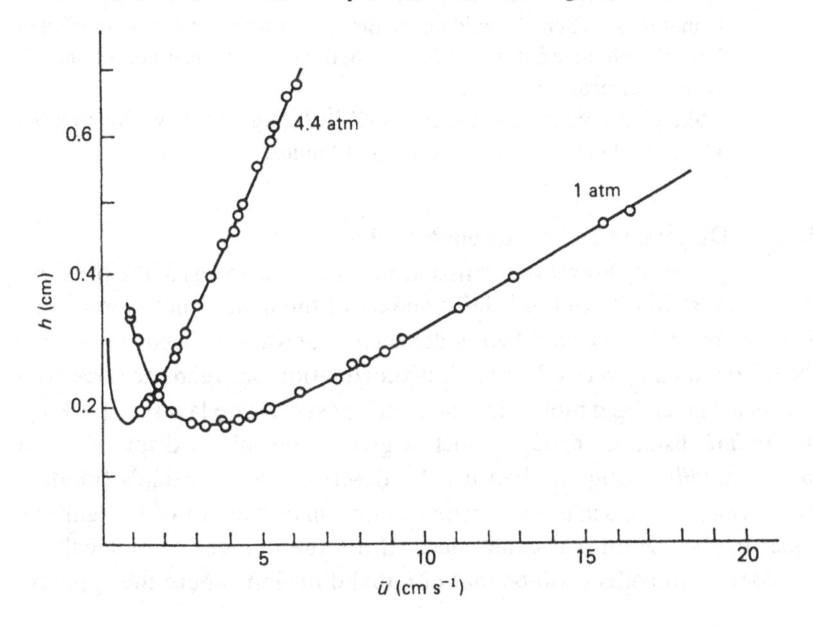

original work [18] showed that again a diffusion-type model was valid. Equation (5.26) would hold, but the appropriate 'diffusion' or dispersion coefficient K would now be given by

$$K = 7a\bar{u}\sqrt{c_f} \simeq 0.6\, a\bar{u}\, Re^{-1/4}, \tag{5.33}$$

in which c_f is the Fanning friction factor. For lower turbulent velocities the dispersion coefficient is found experimentally [22] to be up to ten times as large as indicated by equation (5.33), but as the velocity is raised equation (5.33) is more closely obeyed, the agreement being good when the Reynolds number is greater than about 50 000.

These results are valuable in considering the mixing of fluids pumped in sequence down pipelines, but are not of great consequence in reactor design. In § 5.9*a* we discussed the effect of the velocity profile on the performance of a reactor in laminar flow. Diffusion tends to reduce dispersion in laminar flow, and thus makes the plug-flow model a better design approximation. For turbulent flow, the effect of dispersion is smaller still, and can be neglected in the design of a turbulent-flow tubular reactor, since the effect of temperature variations on the velocity constant is likely to be much more important.

***Example* 5.15**

How do the diffusion coefficient and Schmidt number of a liquid compare with those of a gas?

Two miscible fluids are pumped in sequence down a pipeline of diameter d. When the middle of the intermixed zone has proceeded down the pipeline a distance L, the length l of this zone is approximately $4\sqrt{(KL/\bar{u})}$ provided $l \ll L$.

Sketch graphs on log–log scales of l^2/Ld against Reynolds number where the fluids are (a) gases and (*b*) liquids.

5.10 Dispersion in flow through packed beds

Again the literature on this subject is large, and as in the previous section we shall give only a brief summary of the main conclusions. Fluid flowing through a packed bed is deflected sideways as it passes around the solid packing pieces. If there is no correlation between the directions in which a given fluid molecule is deflected at successive layers of packing, the *radial* distance through which a given molecule is displaced as it moves *axially* along the bed can be described by Einstein's 'random walk' treatment. Each individual deflection will be roughly of a magnitude equal to the size of a packing piece. If a large number of molecules is considered, the effect will be that of radial diffusion, where the apparent diffusion coefficient, D_r, is given by

$$Pe_r = \frac{\bar{u}d}{D_r} \simeq 11. \tag{5.34}$$

Here Pe_r is the *radial Peclet number*, $\bar{u}$ is the average linear velocity, and d is a length characteristic of the packing piece – for a bed of spheres it is the diameter of the spheres. The magnitude of D_r is usually much greater than the *molecular* diffusion coefficient, even for gases, and much greater still for liquids.

A disc of tracer will be dispersed axially as it moves down the bed. The phenomenon can again be described by a diffusion-type model. The magnitude of the apparent *axial*, or *longitudinal*, diffusion coefficient, D_x, is given by

$$Pe_x = \frac{\bar{u}d}{D_x} \simeq 2. \tag{5.35}$$

Here Pe_x is the *axial Peclet number*; $\bar{u}$ and d are the same as in equation (5.34). Reviews of the theory and experimental results are available [**4**, **23**].

Another way of looking at the problem is to regard the interstices between the packing pieces as a chain of mixing tanks. Provided that the number of such tanks in series is greater than about ten, their residence-time distribution function closely resembles that given by the diffusion model (and both are of the form of Fig. 21). The connection between the tanks-in-series model and the diffusion model is that N, the number of tanks in series, is related to D_x by

$$N = \frac{\bar{u}l}{2D_x} = \frac{l}{d} \cdot \frac{\bar{u}d}{2D_x} \simeq \frac{l}{d}. \tag{5.36}$$

In this equation l is the length of the bed and we have used equation (5.35). The answer that the number of tanks equals the length of the bed divided by the particle size gives a very reasonable physical meaning to N.

It must now be asked what effect longitudinal dispersion, with D_x given by equation (5.35), has upon the yield from a packed-bed reactor in the steady state.

A balance on reagent over an elementary length of reactor leads to the differential equation

$$D_x \frac{d^2c}{dx^2} - \bar{u}\frac{dc}{dx} - r = 0. \tag{5.37}$$

In equation (5.37) r is the reaction rate term.

The boundary conditions for equation (5.37) have given rise to considerable discussion [**1, 24**]. It has been shown that the correct boundary conditions are:

$$\left.\begin{aligned} \bar{u}c_0 &= \bar{u}c - D_x\frac{dc}{dx} \quad &\text{at } x=0,\\ \frac{dc}{dx} &= 0 \quad &\text{at } x=l. \end{aligned}\right\} \tag{5.38}$$

Here c_0 is the feed concentration of the reagent species.

Equation (5.37) is non-linear for other than first or zero order reactions. Thus numerical solution is generally required. The following cases and procedures can be noted:

(1) When the reactor is to operate at low conversion only the degree of dispersion or back-mixing is not very important; the difference in volume between a perfectly-mixed reactor and a plug-flow reactor is not very great [**25**].

(2) A diffusion-type model is only really applicable for small deviations from plug-flow; i.e. $D_x/\bar{u}l$ must be small. For a packed bed $D_x/\bar{u}d \simeq \frac{1}{2}$, see equation (5.35), and hence $D_x/\bar{u}l \simeq d/2l$. Provided d/l is less than about 0.1, the diffusion model is applicable, see [**1**]. Graphs are available [**25, 26**] to calculate l/l_0, the ratio of the bed length to the comparable plug-flow bed length. As an example, when $D_x/\bar{u}l = 0.05$, l/l_0 is about 1.1 for a first-order reaction going to 90% completion, and is slightly greater for a second-order reaction.

If the packed bed is longer than 100 packing pieces, and this is nearly always the case, the effect of axial dispersion would be to increase the required length by about 1% over that for a plug-flow reactor. This is a negligible correction in view of uncertainties in kinetics and catalyst reproducibility.

It sometimes occurs that the reaction is so fast that only a very short bed is required. Axial dispersion can then be important. However, in such cases the diffusion model may not be applicable; the tanks-in-series model will be more realistic. In practice channelling, or gross inequalities in the flow rate across the cross-section of the bed, may give rise to serious effects, leading to impermissible quantities of feed reagent getting through the bed unreacted. To avoid this the bed may have to be considerably deeper than first calculated.

The results of some experiments on packed beds of glass or ion-exchange resin beads are instructive [**27**]. Figs. 23 and 24 show F curves for various situations. In each case a step-change experiment was carried out on a bed 1.34 m long packed with spheres about 0.6 mm in diameter. The mean linear velocity, $\bar{u}$, was about 4 mm s^{-1} for all the runs.

With glass beads the F curve was in good agreement with the diffusion model, Fig. 23, apart from some 'tailing'. This has often been reported and shows that the diffusion model does not give a good description of the removal of the last traces of material present in 'dead space' in a packed bed.

With ion-exchange beads, using a step-change involving no exchange reaction with the beads, the results were closely similar to those using

Fig. 23. Step-change experiments. NaCl solution following water.

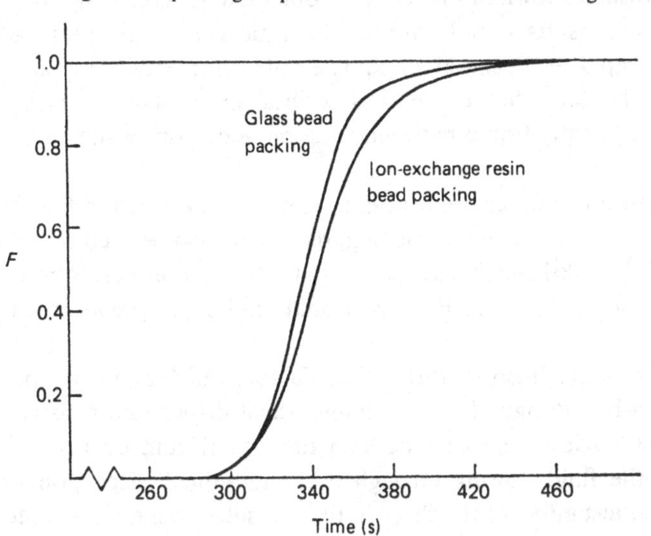

Fig. 24. Step-change experiments. Ion-exchange bead packing. (*a*) NaCl replacing HCl. No dispersion, infinitely fast kinetics. (*b*) NaCl replacing HCl. Experimental curve. (*c*) HCl replacing NaCl. No dispersion, infinitely fast kinetics. (*d*) HCl replacing NaCl. Experimental curve.

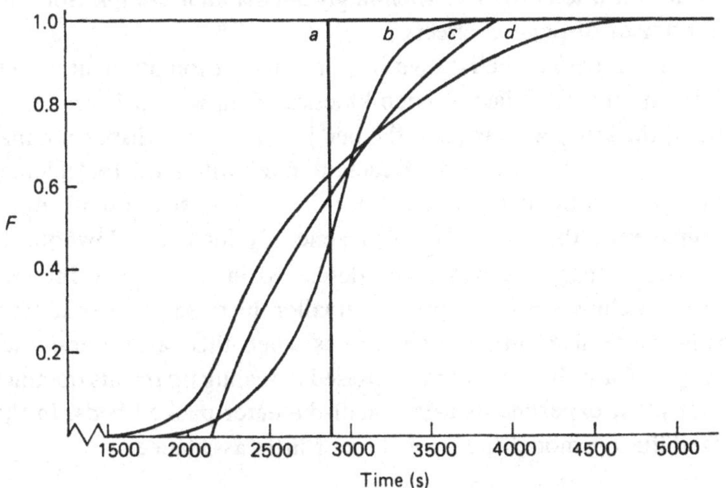

glass beads. The 'tailing' was more marked, and this was probably due to slight absorption of the tracer material within the ion-exchange particles.

With these two runs the value of Pe_x was about $\frac{1}{3}$ in the first case and $\frac{1}{4}$ in the second. These, though lower than the figure of 2 quoted in equation (5.35), are within the range of those found experimentally by many workers [**4**, **23**].

Using a step-change to a solution which reacted with the ion-exchange packing, the results shown in Fig. 24 were obtained. Curves b and d are the experimental results. a and c are the theoretical curves for plug-flow and infinitely-rapid exchange kinetics. The difference between a and c results from the fact that curve a describes an exchange which is 'favoured' by the equilibrium between solution and resin, while curve c refers to an 'unfavoured' exchange.

The breakthrough curves occur at much longer times than in Fig. 23, because of the reaction with the packing, but it can also be seen that the spread of the breakthrough curves is up to thirty minutes, whereas longitudinal dispersion, as in Fig. 23, would lead to a spread of only some two minutes.

These experiments illustrate three of the factors which may be important in packed-bed design. They are longitudinal dispersion, resistance to mass transfer within the packing, and the equilibrium between the packing and the fluid flowing through it. Longitudinal dispersion has clearly the smallest effect of the three in these results. It should be noted that these experiments were approximately isothermal. We have seen, in § 3.6, the very large effect which temperature variations can have on the performance of a packed-bed reactor. It may well be that such temperature variations are overwhelmingly important in comparison with mass-transfer or dispersion effects.

To sum up, in packed-bed design it is unusual for longitudinal dispersion to be an important factor when kinetics or mass-transfer rates are involved. If the latter are fast, and the bed is a short one, dispersion may have to be taken into account. Reactant fluid will tend to 'channel' through regions of the bed where the packing is more than usually open, for example near the walls. This will generally lead to a lowering of reactor performance, and cannot be described in terms of a diffusion model. Channelling is usually worse the smaller the ratio of tube diameter to packing piece size, and can be serious when this ratio is less than about twenty. Care should thus be exercised in scaling up results obtained from pilot-plant experiments using small-diameter packed beds. In this situation diffusion-model theory will be of little assistance.

Example 5.16

For a perfect mixer $I(t) = E(t)$, since the contents have the same age distribution as the outlet stream. Use equations (5.1) and (5.4) to derive equation (5.14).

Example 5.17

A piece of process equipment consists of two well-mixed vessels, A_1 and A_2, each of volume 40 m^3 connected in series. In parallel with these is a plug-flow section, B, connected as shown below. A solution is fed to the system at a rate of 100 $m^3 h^{-1}$. A pulse of tracer is injected at the inlet at $t = 0$. The exit residence-time distribution, $E(t)$, is measured and found to consist of two peaks, a 'broad' one with a maximum, appearing at $t = 40$ min, followed later by a sharp 'spike', appearing at $t = 120$ min.

In a separate experiment the vessel B is replaced by a single well-mixed vessel of volume 80 m^3 which is connected across A_1 and A_2, the flow rates remaining as above. The solution fed under constant-flow conditions is suddenly replaced by a solution containing a tracer at $t' = 0$. The F curve is followed at the system exit. Show that $F(t') = 0.5$ after approximately 1.2 h.

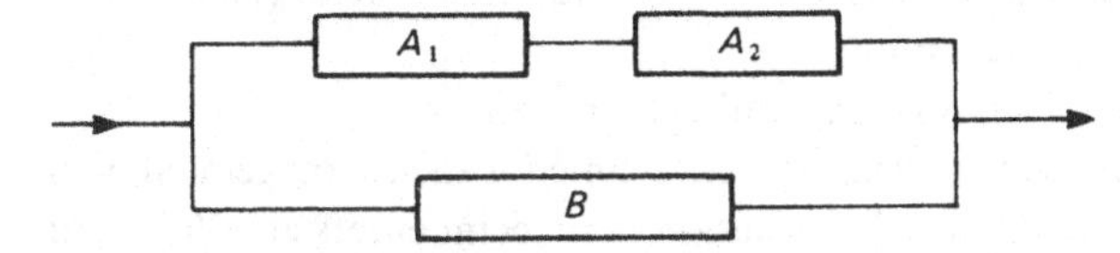

Example 5.18

A laminar flow reactor of circular cross-section is used to study a zero-order reaction. The reaction is irreversible and carried out in solution under isothermal conditions. It is noted that when an inert tracer is introduced at the reactor inlet it first appears at the reactor exit after 1.35 h. Under steady-state reaction conditions a feed containing 1.1 kmol m^{-3} of reactant is reduced to 12% of its initial concentration after passing through the reactor.

Obtain a quadratic equation for the rate constant and hence determine its value. Explain why the other solution to the quadratic is unacceptable. [*Answer.* 0.533 kmol $m^{-3} h^{-1}$.]

Example 5.19

The residence-time distribution of a continuous reactor is found by tracer measurements to be given by

$E(\theta) = e^{-\theta}$, where the dimensionless time is $\theta = vt/V$.

The reactor is to be used to carry out the decomposition of a substance A in solution. This reaction is irreversible and of zero order with velocity constant $k = 0.5$ kmol m^{-3} h^{-1}. If the feed concentration is 2.0 kmol m^{-3} and its flow rate, v, 10 m^3 h^{-1}, use the above residence-time distribution to estimate the concentration of A in the effluent from the reactor, of which the volume V is 30 m^3.

The above residence-time distribution is the same as that of a single perfect mixer. If the reactor really were such a perfect mixer, what would the concentration of A in the effluent be? Why are the two answers different?

[*Answers.* 0.895, 0.5.]

Example 5.20

A first-order liquid phase exothermic reaction is to be carried out to give an 80% conversion to product. The dependence of the reaction rate constant on temperature may be represented over the relevant temperature range as $k = 0.004 + 2.0 \times 10^{-5}(T - 400)$, where T is the absolute temperature and k has dimensions s^{-1}.

Two adiabatic reaction vessels in series are to be used; the conversion at the exit of the first vessel is to be 30%. The possible arrangements are:

(*a*) two tubular reactors;

(*b*) a C.S.T.R. followed by a tubular reactor.

Calculate the overall mean residence-time for each arrangement, and comment on your results. The solution of reactant enters at 400 K, and the initial concentration of reactant is 10.0 kmol m^{-3}. The density and specific heat of the solution are 1000 kg m^{-3} and 201 kJ kg^{-1} K^{-1} respectively and remain constant. Over the relevant temperature range ΔH may be taken as 84 000 kJ kmol^{-1}.

[*Answers.* (*a*) 213.8 s, (*b*) 211.8 s.]

Symbols

a Pipe radius, m.
c Concentration, kmol m^{-3}.
$\bar{c}$ Average concentration, kmol m^{-3}.
c_f Fanning friction factor.
d Pipe diameter, m.
D Diffusion coefficient, m^2 s^{-1}.
E, F, I Residence-time functions, see § 5.2.
H Heaviside unit function.
k Second-order velocity constant, see equation (5.21).

K Apparent dispersion coefficient, see equation (5.31).
L Length down pipe, m.
M Tracer quantity, see equation (5.30).
n, N Number of tanks in series, see § 5.3.
Pe Peclet number, see equations (5.34) and (5.35).
q In § 5.3 amount of tagged substance, kmol.
q In § 5.9 fluid flow quantity, $m^3 s^{-1}$.
r Radial position, m.
Re Reynolds number.
t Time, s.
$\bar{t}$ Mean residence time, s.
u Velocity, $m s^{-1}$.
v Flow rate, $m^3 s^{-1}$.
V System volume, m^3.
x Distance down axis of pipe, m.
β See equation (5.21).
δ Dirac delta function.
θ Dimensionless residence time $= vt/V$.
ψ See equation (5.21).

References

1. Danckwerts, P. V., *Chem. Engng Sci.*, 1953, **2**, 1.
2. Turner, J. C. R., *Chem. Engng Sci.*, 1974, **29**, 1298.
3. Spalding, D. B., *Chem. Engng Sci.*, 1958, **9**, 74.
4. Levenspiel, O. and Bischoff, K. B., *Advances in Chemical Engineering* (Academic Press, New York, 1963), Vol. 4, p. 95.
5. Levenspiel, O. and Turner, J. C. R., *Chem. Engng Sci.*, 1970, **25**, 1605.
6. Turner, J. C. R., *Chem. Engng Sci.*, 1971, **26**, 549.
7. Asbjørnsen, O. A., *Chem. Engng Sci.*, 1961, **14**, 211.
8. Singer, E., Todd, D. B. and Guinn, V. P., *Ind. Eng. Chem.*, 1957, **49**, 11.
9. Todd, D. B. and Wilson, W. B., *Ind. Eng. Chem.*, 1957, **49**, 21.
10. Handlos, A. E., Kunstman, R. W. and Schissler, D. D., *Ind. Eng. Chem.*, 1957, **49**, 25.
11. Danckwerts, P. V., *Chem. Engng Sci.*, 1958, **8**, 93.
12. Zwietering, T. N., *Chem. Engng Sci.*, 1959, **11**, 1.
13. Weinstein, H. and Adler, R. J., *Chem. Engng Sci.*, 1967, **22**, 65.
14. Rys, P., *Angew. Chem.*, 1977, **89**, 847.
15. Villermaux, J. and David, R., *Chem. Engng Comm.*, 1983, **21**, 105.
16. Cleland, F. A. and Wilhelm, R. H., *A.I.Ch.E. Journal*, 1956, **2**, 489.
17. Taylor, Sir Geoffrey, *Proc. Roy. Soc. London*, 1953, **A219**, 186.
18. Taylor, Sir Geoffrey, *Proc. Roy. Soc. London*, 1954, **A223**, 446.
19. Taylor, Sir Geoffrey, *Proc. Roy. Soc. London*, 1954, **A225**, 473.
20. Aris, R., *Proc. Roy. Soc. London*, 1959, **A252**, 538.
21. Evans, E. V. and Kenney, C. N., *Proc. Roy. Soc. London*, 1965, **A284**, 540.
22. Tichacek, L. J., Barkelew, C. H. and Baron, T., *A.I.Ch.E. Journal*, 1957, **3**, 439.

23. Hiby, J. W., Paper C71, *Symposium on the Interaction Between Fluids and Particles*, The Institution of Chemical Engineers, London, June 1962.
24. Bischoff, K. B., *Chem. Engng Sci.*, 1961, **16**, 131.
25. Levenspiel, O. and Bischoff, K. B., *Ind. Eng. Chem.*, 1959, **51**, 1431.
26. Levenspiel, O. and Bischoff, K. B., *Ind. Eng. Chem.*, 1961, **53**, 313.
27. Turner, J. C. R., *Brit. Chem. Engng*, 1964, **9**, 376.

6

Chemical factors affecting the choice of reactor

6.1 Factors affecting choice

In this chapter we shall be concerned with the way in which the nature of the chemical reactions involved in the production of a desired product affects the choice of reactor. It will be shown that one reactor type may be preferable to another because it results in a higher yield, or better quality of product.

Such chemical considerations can have an important bearing on production costs, but they are only one factor in the overall composition of costs. Other factors to be considered include the following:

(i) The capital cost of the process. It is unusual for the capital cost of the reactor itself to dominate in the choice of reactor type. However, the cost of the plant for purifying the product can be high, and such costs can depend significantly on the choice of the reactor producing the product to be purified.

(ii) Running costs of the reactor and related plant. It is clear that raw material costs will be lower the higher the yield, but the cost of labour, and of energy, involved in running the plant may dictate the choice of reactor type. For example, the energy costs of the purification plant may be considerable, particularly if there are large recycle streams to be separated and returned to the reactor.

(iii) Ease of control, and of maintenance. The plant has to provide a product of specified quality, and this will require control equipment. The plant must also be easily started up and shut down. To achieve a small gain in steady-state performance at the cost of increased difficulty in maintaining that steady state can well be uneconomic. Simplicity of design and control possesses real economic benefits.

(iv) Safety. This can be an overriding factor in the choice of reactor type. It is not easy to include safety in the economic calculations, but there has been a rapid growth in recent years of 'hazard and operability' study techniques. These include methods of quantifying safety measures, leading to economic comparisons between alternatives which take safety into account. Such examples of reactor choice being determined by safety considerations arise with nitration or oxidation reactions. These can be dangerously exothermic, and in such cases a C.S.T.R. may be preferred to a tubular reactor because of the greater ease of safe temperature control of the former, even though the latter would have advantages on chemical grounds.

When all such factors are brought into consideration, they may often compel a decision concerning the type of reactor quite different from that which would be arrived at on the basis of chemical factors alone. This chapter will therefore discuss only one aspect, though an important one, of the overall process of choosing a reactor.

Such a choice should, of course, follow an investigation of the optima of all the possible types. However it will be suggested that, as far as chemistry is concerned, it is usually possible to form a sound conception of the most suitable reactor type by using quite simple arguments. The chemical distinctions between the main reactor types can be analysed in terms of:

(*a*) differences of residence-time distribution;
(*b*) differences of concentration history;
(*c*) differences of temperature history.

There are really two problems: (i) how best to design a plant for given kinetic parameters; (ii) how to vary the kinetic parameters to the improvement of the design. We are mainly concerned with the first of these problems in this chapter and hence will be dealing with (*a*) and (*b*). Factor (*c*) above relates to the second of these problems and will be discussed in Chapter 9.

We shall here be discussing batch reactors, tubular reactors, and the C.S.T.R., whereas in practice many important variants, including the fluidized reactor, are to be found in industry. Although this limits the comprehensiveness of the treatment, the consideration of the above-named reactors, which are limiting and idealized types, will be sufficient to bring out the essential factors. These can then be applied to other types of reactors as required.

6.2 Yield and conversion

Before proceeding, it is desirable to define the yield, and conversion, of a reaction. For a reaction

$$A+B\rightarrow X$$

an 80% molar yield means that 0.8 moles of X are obtained for every 1 mole of A or B (whichever is not in excess), the remaining 20% being accounted for either as unwanted by-products or as unchanged A, or B. In the latter case the cause may be either insufficient duration of reaction, or the attainment of an equilibrium state appreciably short of complete conversion. In practice a desired product may be the result of several reactions carried out in sequence, perhaps in separate reactors. In such a case one can refer to the *overall* yield of a process. Finally, yield is sometimes defined as the *weight* of product per unit weight of main reagent. As Example 6.1 shows, a yield so defined can be greater than 100%.

Yield refers to the amount of product made. Conversion, on the other hand, refers to the proportion of a reagent which reacts in the system. If there is only one possible reaction, and there is no recycle of unreacted reagent, the conversion equals the molar yield. If unreacted reagent is recycled, then conversion must be more precisely defined. 'Pass conversion' or 'conversion per pass' means the proportion of the reagent entering the reactor (including both the recycle stream and the 'fresh' feed) which has reacted by the time it reaches the reactor effluent. 'Overall conversion' would refer to the proportion of the fresh feed supplied to the plant which is converted in the streams leaving the plant. The synthesis of ammonia is an important case in which the distinction between pass conversion and overall conversion must be made. Example 6.2 is drawn from this reaction.

Where a given reagent may react in various ways, yield and conversion will not be so simply related, since yield will usually refer to a single, desired product among the many, all of which go to making up the conversion.

In the definition of yield, care has to be taken to allow for the stoichiometry of the reaction, and also a decision must be taken whether the yield should be computed relative to the amount of reagent introduced into the system or relative to the amount of reagent used up. Let the reaction be

$$\alpha A + \beta B = \chi X,$$

where α, β and χ are stoichiometric coefficients, and assume that it is reagent A which is not present in excess.* At the termination of reaction let (X) be the moles of X formed and $(A)_r$ be the moles of A which have reacted. Finally, let $(A)_t$ be the total moles of A which were

* As will be shown below, there are often important economic advantages in using an excess of one or other of the reagents.

originally introduced into the system. The yield might be then defined either as

$$\Phi' = \frac{\alpha}{\chi}\frac{(X)}{(A)_r}, \tag{6.1}$$

or as

$$\Phi = \frac{\alpha}{\chi}\frac{(X)}{(A)_t}. \tag{6.2}$$

In both instances the introduction of the stoichiometric factor α/χ is necessary in order to make the yield equal to unity (or 100%) if reaction to X were complete. Obviously the two definitions are the same if *all* A reacts, but equation (6.2) is the more appropriate if, at the end of reaction, there remains some unreacted reagent which is not recoverable. Conversely equation (6.1) might be the economically more significant of the two definitions if the unreacted reagent were recoverable.

***Example* 6.1**

Benzene is reacted with propylene to produce cumene in virtually 100% molar yield. This is then used to make phenol and acetone in a two-step process, involving oxidation to a hydroperoxide, followed by 'cleavage' to yield the two main products. The oxidation produces a molar yield of 91% hydroperoxide, the rest going to heavy by-products of essentially the same molecular weight as the hydroperoxide. The hydroperoxide is separated, and the cleavage reaction is 100% effective.

85 000 tonnes per year of phenol are to be produced. What will be the production of acetone, and of heavy by-products? What are the overall molar conversion, and weight conversion, of benzene to phenol?

[*Answers.* 52 450; 13 590 tonnes yr^{-1}; 91%, 110%.]

***Example* 6.2**

Ammonia is synthesised from a stoichiometric mixture of nitrogen and hydrogen, which also contains 1 mol % of argon. The stream leaving the reactor goes to a separator, where ammonia product is removed. The unreacted gases are recycled to the reactor, but a purge is bled off from the recycle stream to remove argon from the system. This purge stream amounts to 5% of the original synthesis gas (molar ratio). The pass conversion in the reactor is 15%. What is the concentration of argon in the recycle stream? How large is the recycle stream fed to the reactor, per mole of fresh feed, and what is the overall conversion to ammonia?

[*Answers.* 20%, 6.68, 96%.]

***Example* 6.3**

A substance, A, is converted to a product, P, in a C.S.T.R. of volume $V\ m^3$. The feed to the plant, $v\ m^3\ h^{-1}$, is a dilute solution of A, c_{A0} kmol m^{-3}, and the reaction is first order in A with velocity constant $k\ h^{-1}$.

The cost of the process is proportional to the volume V of the reactor (which has to be made of an expensive material). The income of the process is directly proportional to the amount of P made. Derive an equation to determine the volume of reactor which will maximize the profit from the system, for given values of v, c_{A0} and k. (In this example the unit of time is the hour.)

Solution

The material balance and reaction equations lead to

$$v(c_{A0}-c_A)=kVc_A \tag{6.3}$$

in the steady state. Hence

$$c_A=\frac{c_{A0}}{1+\dfrac{kV}{v}}. \tag{6.4}$$

The income of the process is proportional to the amount of P made. Since A reacts to give P only, it follows that

$$\text{Income per hour}=av(c_{A0}-c_A). \tag{6.5}$$

It is also known that

$$\text{Cost per hour}=bV,$$

where a, b are proportionality constants. Hence

$$\text{Profit per hour}=av(c_{A0}-c_A)-bV. \tag{6.6}$$

Substituting equation (6.4) for c_A,

$$\text{Profit per hour}=avc_{A0}\left(1-\frac{1}{1+\dfrac{kV}{v}}\right)-bV$$

$$=\frac{akVc_{A0}}{1+\dfrac{kV}{v}}-bV. \tag{6.7}$$

The maximum profit can be found by differentiating equation (6.7) with respect to the variable V. Putting this derivative equal to zero, and rearranging, we obtain

$$V=v\left[\sqrt{\left(\frac{ac_{A0}}{bk}\right)}-\frac{1}{k}\right]. \tag{6.8}$$

Example 6.4

If in Example 6.3 the parameters have the following values, calculate the yield which gives the maximum profit, based on the A fed to the plant.

$v = 10\ \mathrm{m^3\,h^{-1}}$,
$c_{A0} = 2\ \mathrm{kmol\,m^{-3}}$,
$k = 25\ \mathrm{h^{-1}}$,
$a = £1\ \mathrm{kmol^{-1}}$,
$b = £2\ \mathrm{m^{-3}\,h^{-1}}$.

[*Answer.* 80%. What is the yield based on the A reacted?]

Example 6.5

The design of Example 6.3 is to be modified. The effluent from the reactor will pass to a separator, where unreacted A is separated as a solid, half of which is suitable for re-use and is fed back to the reactor. For the same values of the parameters as in Example 6.4, what will now be the yield giving the maximum profit?

[*Answer.* 86%.]

6.3 Selectivity and reactivity

Where a given reagent can react to give several different products, the designer has to consider both the total amount of reaction and that proportion of it which goes to produce the desired product. In practice one can generally alter operating conditions so as to allow for small errors in *sizing* a reactor. The reactor may not produce quite as much as originally hoped for, but it can be operated profitably. But if the reactor produces the wrong product, this defect is not so easy to remedy.

The 'selectivity', when referring a reactor (or to a catalyst packing for a reactor), is a measure of that proportion of the total reaction which produces the desired product. Different bases can be used; for example, one definition of selectivity would be the ratio of reagent reacting to give the desired product to that giving the undesired products. It may well be preferable to choose a *selective* catalyst of low *reactivity* (leading to a large reactor), rather than a highly *reactive* catalyst of low *selectivity.* The latter would result in a smaller, cheaper reactor, but might lead to unacceptable costs of raw materials.

For a proper calculation, the costs of feedstock and the value of the different products made must be considered, together with the costs of building and running both the reactor and the equipment necessary to separate the products.

***Example* 6.6**

A reagent A costs £20 per tonne (one tonne = 1000 kg). It can react in two ways. By the first reaction one tonne of A produces 1.2 tonne of product X, worth £x per tonne. By the second reaction one tonne of A produces 0.8 tonne of Y, worth £y per tonne. Defining the selectivity as the ratio of A consumed to give X to that consumed to give Y, estimate the minimum selectivity required for the process to be considered as a viable proposition – for the following values of x and y.

(a) $x = 100, \quad y = 5$;
(b) $x = 30, \quad y = 5$;
(c) $x = 20, \quad y = 5$.

[*Answers.* (a) 0.16, (b) 1, (c) 4.]

***Example* 6.7**

Comment on the answers of Example 6.6 and their implications in the design of chemical reactors.

6.4 Consecutive reactions

We shall consider reactions of two main types:

(a) $A \rightarrow B \rightarrow C(\rightarrow D \rightarrow)$,

where each step either involves no other reagent, or involves another reagent present in considerable excess, so that there is little competition between A, B, C, etc., for this other reagent;

(b) $M + N \rightarrow P$

$P + N \rightarrow Q$

$Q + N \rightarrow R$, and so forth,

where the reagent N is not in great excess, and M, P, Q, etc., are in competition for what is available.

An example of type (a) is the thermal cracking or degradation of hydrocarbons; the aim of the designer is to 'stop' the reaction when there is maximum yield of some intermediate or intermediates (e.g. B or C). Many oxidation reactions are of this type, too. Methanol can be oxidized to formaldehyde, and this will be oxidized further to CO_2 unless the reaction is suitably controlled.

Examples of type (b) occur in organic substitution reactions. If the aim is to make monochlorbenzene (or nitrobenzene), then it would be desirable to use roughly stoichiometric quantities of the reagents. But the first monochlorbenzene that is made will compete with benzene for chlorine, and, if successful, will produce dichlorbenzene, probably an undesired product.

Returning to reactions of type (*a*), if we consider a batch reactor, initially supplied with A, and B is the desired product, it is clear that the proportion of B passes through a maximum at some optimum time, t_m. This is indicated in Fig. 25. Hence to obtain the maximum yield of B, relative to the amount of A introduced to the system, the reaction process should have a duration of t_m, no longer and no shorter. This condition is best satisfied by using a batch reactor, although this may be undesirable on other grounds.

For any type of continuous reactor there is inevitably a distribution of residence-times. Thus even though the mean residence-time may be put equal to t_m, there will be elements of fluid passing through the system with residence times both greater and less than this 'optimum' value. We have seen in the last chapter that the residence-time distribution is not in general sufficient to determine conversions uniquely, but equation (5.20) may be used to give an approximate indication of the effect of a spread of residence times. Such a spread will generally reduce the maximum obtainable yield of an intermediate, B.

Fig. 25. Degradation reaction: relative yields obtainable by batch and by C.S.T.R. (as worked out for $k_2/k_1 = 2.0$).

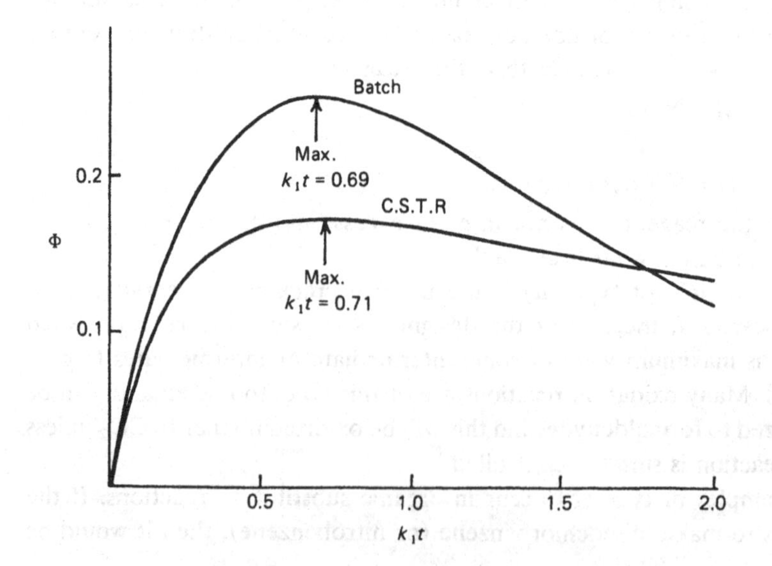

A tubular reactor in which there is a fairly good approximation to plug flow will clearly give a yield not far below the batch yield and such reactors are actually used for many reactions of this type, e.g. the oxidation of methanol on silver catalyst.

The lowest yield may be expected to be given by a single-tank C.S.T.R. and it is instructive to show in detail that it can be very much lower than is obtainable by a batch process. The reaction sequence

$$A+X \rightarrow B, \qquad B+X \rightarrow C,$$

will be discussed and it will be supposed, for simplicity, that the rates of the reactions are proportional to the concentrations of A and B, respectively, as might be the case if the concentration of the other reagent X were in large excess. A reaction scheme of this kind has been examined experimentally [**1**] and has been discussed by other authors [**2, 3**].

Assuming negligible volume change in the reaction, we have for batch or plug-flow conditions

$$\frac{dc_A}{dt} = -k_1 c_A, \tag{6.9}$$

$$\frac{dc_B}{dt} = k_1 c_A - k_2 c_B, \tag{6.10}$$

where c_A and c_B are the concentrations of A and B respectively.*

We have previously solved a problem closely resembling this – Example 2.3 – from which we deduce

$$\left[\frac{c_B}{c_{A0}}\right]_{\max} = s^{[s/(1-s)]}, \tag{6.11}$$

where $s = k_2/k_1$, and c_{A0} is the initial concentration of A.

We shall assume that the economically significant yield of B is that which is computed relative to the amount of reagent introduced into the system. Hence, from the defining equation (6.2),

$$\Phi = c_B/c_{A0}$$

and its maximum value, $\Phi_{m,B}$, is given by equation (6.11).

Fig. 5 shows that, when $s=0.5$, $\Phi_{m,B}=0.5$ at $k_1 t_m = 1.38$. Fig. 25 shows a similar curve for $s=2.0$. The maximum now occurs at $k_1 t_m =$ 0.69, and $\Phi_{m,B}=0.25$, a lower value. This result is not surprising, since the degradation of B is comparatively more rapid in the second case; as s increases, so $\Phi_{m,B}$ decreases.

* The quantities k_1 and k_2 are not necessarily 'pure' velocity constants; they may be the products of such constants with the concentration (or some function of the concentration) of the other reagent X, this concentration having been assumed constant.

For the same reaction, carried out in a single C.S.T.R., the equations are

$$v(c_{A0}-c_A)=k_1 Vc_A, \tag{6.12}$$

$$vc_B=k_1 Vc_A-Vc_B. \tag{6.13}$$

These are easily solved, to give

$$\frac{c_A}{c_{A0}}=\frac{1}{1+k_1\bar{t}} \tag{6.14}$$

and

$$\Phi_C=\frac{k_1\bar{t}}{(1+k_1\bar{t})(1+k_2\bar{t})}, \tag{6.15}$$

where $\bar{t}=V/v$. This yield has a maximum value at a holding time $\bar{t}_m$ given by

$$\bar{t}_m=(k_1k_2)^{-1/2}. \tag{6.16}$$

Hence substituting this expression in equation (6.15) we obtain the following expression for the highest yield obtainable from a single-stage C.S.T.R.:

$$\Phi_{m,C}=(1+s+2s^{1/2})^{-1}. \tag{6.17}$$

Finally, we can take the ratio of this maximum to that previously derived for the batch or plug-flow conditions:

$$\frac{\Phi_{m,C}}{\Phi_{m,B}}=\frac{(1+s+2s^{1/2})^{-1}}{s^{[s/(1-s)]}}. \tag{6.18}$$

This ratio is less than 1, and is smallest when s is unity, i.e. when the velocity constants k_1 and k_2 are equal, as is sometimes approximately the case for reactions of successive substitution. If the ratio is evaluated for $s=1$ (taking care concerning the value of the exponential expression

Fig. 26. The ratio $\Phi_{m,C}/\Phi_{m,B}$ of equation (6.18) for various values of s.

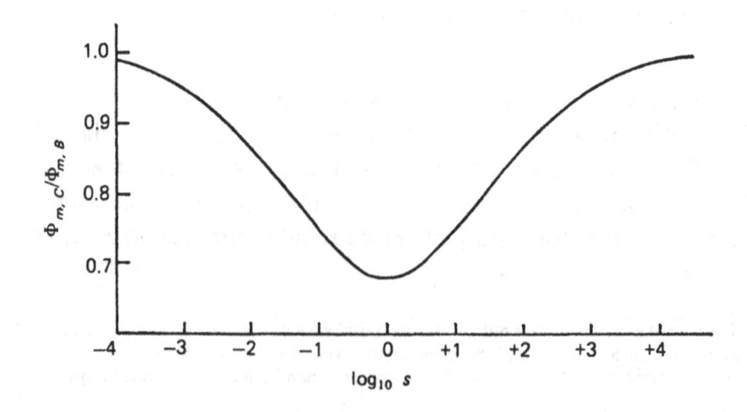

at the limit!) it may be shown that the C.S.T.R. will give only 68% of the yield obtainable by the batch process. This is evidently a large effect which could have an important influence on the economics of the process. Values of the ratio for other values of s are shown in Fig. 26.

As was remarked in § 6.1, certain quite different considerations may make it desirable to use a C.S.T.R. despite its lower yield. If this be the case, the yield could be raised – and raised considerably – by using more than one tank in sequence. Moreover, the opportunity would then offer itself of choosing the relative capacities of these tanks in an optimum manner. In examples where the useful and the degradative reactions are both first order, the capacities should be chosen as equal [**4**]. On the other hand, if the former reaction is of the higher order, the capacities should become larger from the first tank onwards [**4**], and conversely.

If the designer is faced with a problem involving consecutive reactions, what can he do to improve the yield of the desired product? If the velocity constants of the different reactions are known, then in principle any kinetic scheme can be worked out. It is only in a few cases that analytical solutions of general application can be obtained; numerical methods for specific values of parameters are usually necessary. The computer can be used to give valuable and rapid information on the effect of a change of parameters.

The theory of the degradative reaction has been extended in relation to the influence of longitudinal diffusion in a tubular reactor [**5–7**]. Whereas in simple reactions this type of diffusion is not usually very significant, in degradative reactions it can have a considerable influence on the maximum attainable yield.

Other theoretical papers of interest deal with geometrical interpretations of reaction paths [**8, 9**], and consecutive reactions of various orders [**10**]. An example which has been studied experimentally is [**11**] the formation of furfural from pentose. The furfural so formed tends to degrade, and various methods of increasing the yield of furfural were examined.

For consecutive reactions of type (b) the degree of *micro*mixing may be important, and in that case the residence-time distribution is not sufficient to determine the reactor product. As an example, consider the reaction of a droplet of M which is suspended in a solution of reagent N. At the outer surface of the droplet the reaction

$$M + N \rightarrow P$$

may take place. For further M to react, N must diffuse towards the centre of the droplet, and this N will first encounter P, which may lead to

$$P+N \rightarrow Q$$

It can be seen that this may result in a high conversion of M to Q, with only small amounts of P being present at any time. This can happen even if M is miscible with the solution of N, for if the reaction is fast enough there may not be enough time to mix, or disperse, the M throughout the solution before it reacts. In such a case 'packets' of M are converted to Q, without it being possible to stop the reaction at P. This will be the situation if the time for the majority of reaction to occur is smaller than the time for mixing or diffusion to spread the M throughout the solution, and the latter is a sensitive function of the degree of stirring. Both the yields, and the chemical kinetics, may be affected, and Rys **[12]**, for example, discusses the phenomenon of 'mixing-disguised chemical reactions in solution'.

Provided mixing times are short compared with reaction times, the kinetic equations can be integrated, if necessary by computer, as described earlier for reactions of type (a), but in this case the concentration of N will fall with time since it is not in great excess, and we will not be able to use the 'pseudo'-velocity constants referred to in equation (6.10).

***Example* 6.8**

Show that equation (6.15) can be obtained by equation (5.20) in conjunction with equations (6.9), (6.10) and Example 2.3. Why does equation (5.20) yield the correct result in this case?

***Example* 6.9**

Evaluate equation (6.18) for the case $s=1$ and thus show that, for the conditions specified, the C.S.T.R. will give only 68% of the yield obtainable by a batch process.

6.5 Polymerization reactions

Whereas in the previous section we were concerned with the influence of reactor type on reaction yield, we shall here be concerned with its influence on product quality. It will be shown, in the case of polymerization reactions, that the use of different kinds of reactor may result in quite large differences in the molecular weight distribution of the polymer produced.

Tubular reactors can be used for gaseous polymerizations (e.g. high-pressure polyethylene). For liquid-phase, or emulsion polymerization the high viscosity of the liquid, or the necessity for agitation, rule this out. The velocity profile in a tubular reactor due to high viscosity would result in a broad residence-time distribution and one of the consequences would be an insufficient degree of polymerization in material flowing close to the tube axis and an excessive degree of polymerization in material flowing close to the wall. This would tend therefore to the deposition of solid polymer on the wall, resulting in a progressive choking of the tube.

The main alternatives to be considered are therefore batchwise reaction and the use of the C.S.T.R. In the latter instance it is usually desirable to use several vessels in series, but certain polymerizations (styrene, tetrafluorethylene) are so fast that a single-tank C.S.T.R. may suffice.

The decision between batch operation and the C.S.T.R. depends, of course, on a number of factors and one of the most important is scale of output – large scale favouring, as always, a continuous type of process. However, another important factor is the influence of the reactor on the required type of product. Plastics are never single chemical substances but are mixtures of substances having the same general structure but different molecular weight. This arises quite naturally from the intrinsic probability factors in the reaction itself; not every molecule becomes 'activated', or undergoes a suitable collision, at the same instant and thus individual polymer molecules grow to quite different chain lengths. In fact if M stands for monomer and P_i for the ith polymer we have a reaction sequence of the kind

$$M+M \rightarrow P_2, \qquad P_2+M \rightarrow P_3, \quad \text{etc.}$$

It follows that a given sample of polymer is characterized by the distribution of chain lengths about the mean, as well as by the mean itself. The point to be brought out is that the breadth of this distribution depends on whether a batch system or a C.S.T.R. is used for producing the polymer. Since the breadth has an important influence on the mechanical and other properties of a plastic, this can be an important factor influencing the choice of process.

When the mathematics of polymerization kinetics are examined [**13, 14**] it is found that two opposing factors have an influence on the molecular weight distribution. These are:

(*a*) the residence-time, this being equal for all molecules in a batch process, but greatly varying in the case of a C.S.T.R.;

(*b*) the concentration history, and in particular the fact that the monomer concentration diminishes in a batch process but remains constant in each vessel of a C.S.T.R.

As regards the first of these it is intuitively fairly obvious that any spread in the residence times of individual molecules tends to increase the spread of the molecular weights. Some of the growing polymer molecules 'escape' into the outlet from the C.S.T.R. after a very short time and thus do not succeed in growing to an appreciable chain length. Others remain in the vessel for a long time and consequently have the opportunity of reaching a very high molecular weight.

The influence of the second factor is less obvious. However, the significant point is that, in a C.S.T.R., the monomer concentration is stationary and is at a lower average level than it would be, for the same feed conditions, if the same reaction were carried out batchwise. The effect of this factor, in many types of polymerization kinetics, is to reduce the variation in molecular weights.

Which of the two factors is the dominant one depends on the type of polymerization reaction. If the latter involves no processes of termination, as in polycondensation, it is the first factor which dominates and this results in a broader distribution of molecular weights from a C.S.T.R. than would be obtained from a batch process. Such will be the case in the polymerization of monomers such as

$$HO-(CH_2)_n-COOH,$$

where growth occurs by a process of successive esterification with elimination of H_2O:

$$HO-(CH_2)_n-COOH+HO-(CH_2)_n-COOH$$
$$\rightarrow HO-(CH_2)_n-COO-(CH_2)_n-COOH, \text{ etc.}$$

Fig. 27. Polymerization: case where the life of the active polymer is long, or where there is no termination reaction.

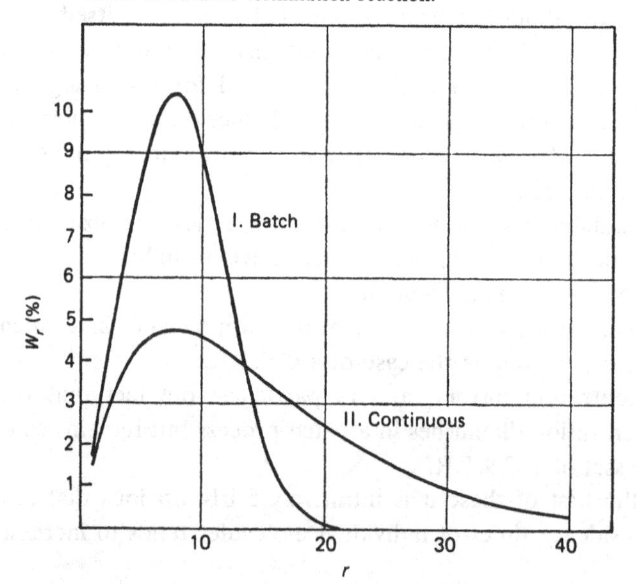

Some calculated distribution curves are shown in Fig. 27, where W_r is the weight percentage of molecules in the total product that are a multiple r of the chain length of the monomer.

On the other hand, in polymerizations which take place by mechanisms involving free radicals or ions,* the life of these actively growing centres may be extremely short owing to the occurrence of termination processes such as the union of two free radicals. If the average life of these centres is very much smaller than the average residence time in the C.S.T.R., the factor (*a*) above is of very little significance and factor (*b*) tends to dominate. Under these conditions it can be shown that, for many types of initiation and termination mechanism, the molecular weight distribution will be narrower than would be obtained from a batchwise process. Fig. 28 shows some calculated curves, again from reference [**13**].

This factor is known to be important in the manufacture of HDPE (high-density polyethylene). This is carried out at *low* pressure, using a stereospecific catalyst of the Ziegler type. It is also important in the emulsion polymerization of styrene [**15**].

More recently, Villermaux [**16**] has shown how variation in the degree of segregation, or micromixing, in a continuous polymerizer can affect both the molecular weight distribution, and the degree of branching of the polymer.

Example 6.10

A monomer is polymerized in a C.S.T.R. There is no change of volume due to the reaction. Monomer reacts with monomer and with polymer molecules, but polymer molecules do not react with each other. All the reactions are second order with the same velocity constant k. Derive equations giving the weight fraction of polymer containing r monomer molecules.

Solution

The C.S.T.R. mass balance on the monomer can be written

$$v(m_0 - m) = km^2V + kmV\sum_{2}^{\infty} p_r{}^{\dagger} \qquad (i)$$

* A typical 'initiation' step, giving rise to a free radical, may be written

$$CH_2{=}CHX \rightarrow \overset{|}{C}H_2{-}\overset{|}{C}HX.$$

A subsequent 'propagation' step, involving growth of the active centre would be

$$\overset{|}{C}H_2{-}\overset{|}{C}HX + CH_2{=}CHX \rightarrow \overset{|}{C}H_2{-}CHX{-}CH_2{-}\overset{|}{C}HX.$$

A 'termination step' might be one involving the union of two such free radicals thus stopping subsequent growth, or it might involve disproportionation, etc.

† The reader should consider why the first term on the right-hand side of (*i*) is km^2V instead of $2km^2V$. A related problem is that of collision between like and unlike molecules in the kinetic theory of gases; 'double counting' must be avoided. For similar reasons there is a $\frac{1}{2}$ factor in (*iii*).

where V is the volume of the tank, v is the flow rate, m_0 is the feed concentration of monomer, m is the concentration of monomer in the tank, and p_r is the concentration of polymer, containing r monomer molecules. This equation can be written

$$m_0 - m = km\bar{t}\left(m + \sum_2^{\infty} p_r\right), \qquad (ii)$$

where $\bar{t} = V/v$, the mean residence-time. Similar balances on the dimer, trimer, etc., give

$$p_2 = km\bar{t}(\tfrac{1}{2}m - p_2), \qquad (iii)$$

$$p_3 = km\bar{t}(p_2 - p_3), \quad \text{etc.} \qquad (iv)$$

Summing,

$$\sum_2^{\infty} p_r = \tfrac{1}{2}k\bar{t}m^2. \qquad (v)$$

Fig. 28. Polymerization: case where the life of the active polymer is short. I, Batch reaction. II and III, continuous reaction.

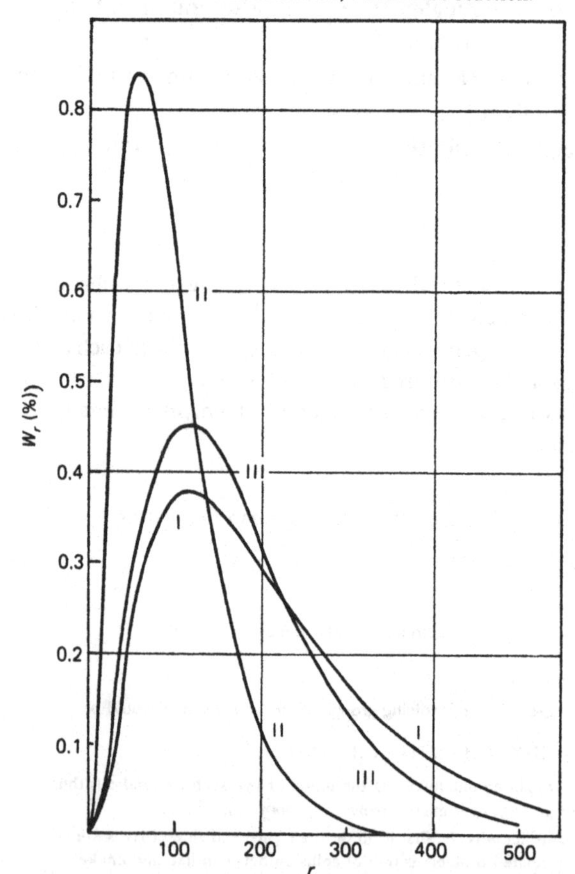

Also, solving successively for p_2, p_3, we can deduce

$$p_r = \tfrac{1}{2}m/[1+(km\bar{t})^{-1}]^{r-1}. \qquad (vi)$$

Now the weight fraction of polymer containing r monomer molecules is

$$W_r = \frac{rp_r}{m_0 - m} = \frac{\frac{1}{2}rm}{m_0 - m}\cdot[1+(km\bar{t})^{-1}]^{1-r}. \qquad (vii)$$

This last equation is what is required, but it can be rewritten in terms of $x = m/m_0$, the fraction of monomer remaining unreacted, and $y = km_0\bar{t}$, which can be regarded as a dimensionless reaction group. We obtain

$$W_r = \frac{r}{\frac{2}{x}-2}\cdot\left(1+\frac{1}{xy}\right)^{1-r}. \qquad (viii)$$

To be able to calculate W_r we must know x, which is itself a function of y, the independent variable. If we substitute from equation (v) in equation (ii) we obtain

$$m_0 - m = k\bar{t}m^2 + \tfrac{1}{2}k^2\bar{t}^2m^3,$$

which in terms of x and y becomes

$$1 - x = yx^2 + \tfrac{1}{2}y^2x^3. \qquad (ix)$$

This last enables x, and hence W_r, to be calculated for any given value of y.

Example 6.11

If, in Example 6.10, $y = 10$, for which value of r is W_r the greatest, and what is this greatest value of W_r?

[*Answers.* 2 or 3, 0.1667.]

Example 6.12

If $y = 100$, for which value of r is W_r the greatest, and what is the greatest value of W_r? Note that y has to be large for a maximum at $r \neq 2$ to be possible. Comment on what you would expect a *batch* reactor to produce.

[*Answers.* 6, 0.0671.]

All these answers are specific to this particular simple reaction scheme: actual polymerization schemes may be much more complicated.

6.6 Crystallization

The process of crystallization, where each existing crystal becomes larger by the stepwise deposition of solute molecules on its surface, is formally very similar to the polycondensation process which has just been discussed. The analogue of the molecular weight distribution of the polymers is the size distribution of the crystals. It may be expected therefore that crystallization will be characterized by important differences of size distribution according to whether it is carried out in a C.S.T.R., on the one hand, or in a batch or tubular system, on the other.

Consider, for example, crystallization taking place in a C.S.T.R. type of system where feed solution enters the first tank in series and each tank is sufficiently well stirred for the crystals to be maintained in suspension. In each tank two processes may be taking place: (*a*) the formation of new nuclei; (*b*) the growth of the nuclei to form crystals of appreciable size. It follows that the crystalline material in suspension in the liquid leaving the ith tank is composed of (*a*) new crystals which came into existence in this tank and (*b*) old crystals which came into existence in earlier tanks and have merely become larger in the ith. By consideration of these factors, and by use of appropriate expressions for the nucleation and growth rates, it is possible to develop the theory of the size distribution in the resulting crystalline product [**17, 18**].

The converse process of the dissolution of crystals can be similarly examined [**19**], as can that of reaction and/or attrition of solid particles in fluidized beds [**20**].

Example 6.13

A hot solution of substance X in water flows continuously into a well-stirred tank which is provided with cooling so that X crystallizes. The stirring is vigorous and the crystalline particles are fairly small, with the result that the concentration of crystal in suspension is uniform throughout the volume of the magma and in the outflow from the tank. Stationary conditions of temperature and supersaturation are maintained. The process of nucleation is spontaneous and its rate depends only on the supersaturation and temperature. The rate of growth of the crystals, which may be taken as being approximately spherical, likewise depends only on the supersaturation and temperature; in particular the linear growth rate of a crystal normal to its surface is independent of its size.

If the aggregation and fracture of crystals may be neglected, show that the number fraction of crystals in the outflow whose radius lies between R and $R+dR$ is equal to $a\,e^{-aR}\,dR$, where a is a constant. Show also that the weight fraction of crystals in the outflow whose radius lies between the values zero and R' is given by

$$\int_0^{R'} R^3\,e^{-aR}\,dR \Big/ \int_0^{\infty} R^3\,e^{-aR}\,dR.$$

6.7 Parallel reactions

The occurrence of parallel reactions is quite common in organic chemical production and is one of the important causes of poor yield of the desired substance. For example, in processes for making meta-substituted benzene compounds the yield is almost invariably less than 100% owing to the simultaneous production of ortho- and para-substituted material. In other instances the side reactions may be such as result in the production of tarry products or CO_2, etc.

If it occurs that the desired and the undesired reactions differ in their kinetic orders there is an important opportunity for making a rational choice between the different reactor types [**3**, **21**]. This point turns on the fact that a difference of kinetic orders implies a difference in the influence of concentration on the relative rates of the reactions. It follows that a C.S.T.R., which gives rise to a concentration history different from that of a batchwise or tubular reaction process, may result in either a higher or a lower yield of useful product, according to the circumstances of the reaction.

Consider two competing reactions which, without prejudice to their actual stoichiometry or mechanism, will be written symbolically

$$A+B \rightarrow X, \qquad A+B \rightarrow Y,$$

where X is the desired product. Let it be supposed that the rates r_x and r_y of the two reactions are proportional to functions $f(a, b, x)$ and $f(a, b, y)$ of the concentrations respectively. Thus

$$r_x = k_1 f(a, b, x), \qquad r_y = k_2 f(a, b, y).$$

Then the ratio of the amount of X to the amount of Y formed in an infinitesimal period of time is equal to the ratio

$$\frac{r_x}{r_y} = \frac{k_1 f(a, b, x)}{k_2 f(a, b, y)}. \tag{6.19}$$

It follows that the reaction conditions should be chosen in such a way that this ratio (which can be called the 'instantaneous selectivity', see § 6.3) always has its highest value.

For example, it might occur that the rate of the first reaction is equal to $k_1 a^2 b$, and the rate of the second reaction to $k_2 ab$. The above ratio would therefore have the value $k_1 a/k_2$. This implies that the yield of X will be greater the higher is the concentration of reagent A, and furthermore that the conditions will become less favourable as reaction proceeds due to the fall in reagent concentration.

Generalizing from this example, it will be clear that when the useful reaction is of higher order than the wasteful one, good yield is favoured by raising the reagent concentration. Conversely if the useful reaction is of lower order, the yield is favoured by lowering this concentration, and this may be a useful effect which more than offsets the resulting diminution of reaction rate.

In the former case the batch or tubular reaction processes are evidently to be preferred since they operate at higher average concentrations for the same feed conditions. If, for quite different reasons, there are grounds for choosing the C.S.T.R., the yield can be raised, – though not as high as the batch yield – by increasing the number of tanks in sequence; also, for a fixed number of such tanks, by making their capacities progressively larger from the first onwards, as indicated in Fig. 29(*a*) [**3**, **21**].

In the converse case where the useful reaction is of the lower order, the concentration of reagent should be as low as possible. In many examples this would be achieved by the obvious step of reducing the concentration of the feed solution. However, in other examples certain considerations (e.g. the cost of recovery of solvent) might militate against such a simple method. Under these circumstances the use of stirred tanks would have advantages over a batch or tubular reactor. In such a case the reagent concentration is never high, especially if only a small number of tanks in series is used and if the first is made relatively large. Alternatively much the same effect could be obtained by operating a type of process (batch, tubular or C.S.T.R.) in which successive small increments of the reagent A are added to the other reagent at a rate small compared to the reaction rate, so that the concentration of A is kept at a low level [**3**, **21–23**]. This is indicated in Fig. 30.

It may occur that one of the parallel reactions is reversible. An example occurs in methanol synthesis:

$$CO + 2H_2 \rightleftharpoons CH_3OH$$

$$CO + 3H_2 \rightarrow CH_4 + H_2O.$$

Here if reaction is allowed to proceed too long, methanol formed in the early stages will tend to decompose back to CO and H_2 when the second irreversible reaction forming CH_4 has reduced the concentrations of CO and H_2 to low levels. Given the kinetics of these parallel reactions

Fig. 29. Parallel reactions. Optimum choice of sizes with fixed number of tanks. (*a*) Useful reaction is of higher order. (*b*) Useful reaction is of lower order.

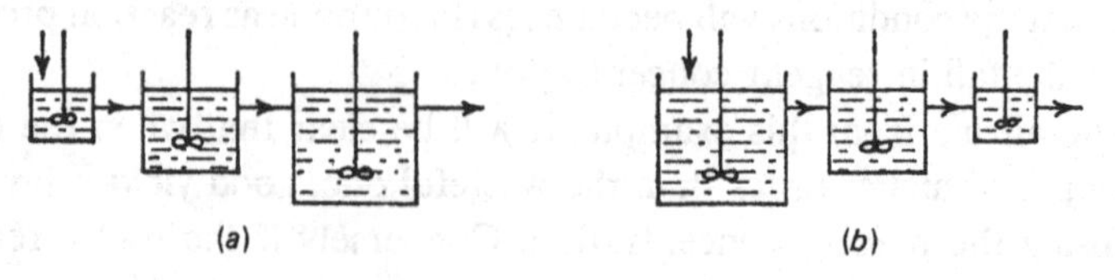

the designer can choose conditions to maximize the production of methanol.

A further example is that of xylene isomerization, which is accompanied by disproportionation reactions. The reactions between the three xylene isomers can be carried out in the gas phase over a catalyst, when they are reversible and first order. Disproportionation reactions also occur and they are effectively irreversible and second order. Thus if isomerization is desired, the pressure should be kept low, and this has consequences in plant design (e.g. a large pressure drop through a packed bed of catalyst is undesirable since it leads to high pressures at the beginning of the bed).

Example 6.14

A substance A is dissolved in a liquid in which it undergoes two reactions at temperature T.

(i) $A \rightarrow B$, isomerization. This is first order in A and has velocity constant k_1;

(ii) $A + A \rightarrow C$, dimerization. This is second order in A and has velocity constant k_2.

The feed solution contains A only, at concentration c_{A0}. Show that the maximum yield of B from an isothermal tubular reactor at T is given by $\beta^{-1} \ln(1+\beta)$, where $\beta = k_2 c_{A0}/k_1$.

Example 6.15

Write down the equations for the reaction of Example 6.14 when it is carried out in a stirred tank. Show that the yield of B can always be made greater than in Example 6.14. What would be the disadvantage of using a stirred tank?

Fig. 30. Maintenance of a low concentration of reagent A by stepwise addition to tubular reactor or chain of stirred tanks.

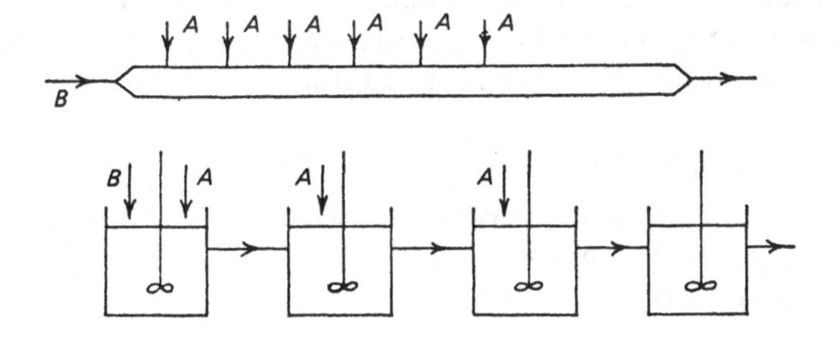

6.8 Parallel-consecutive reactions, coupled reactions

The ideas of the previous sections can be extended to more complicated reaction schemes, involving reversible reactions, or combinations of parallel and consecutive reactions. Two examples can be referred to. First, consider the case [24]

$$A \rightarrow B \rightarrow C, \qquad A + A \rightarrow D.$$

If B is the desired product the consecutive reactions would indicate the use of a tubular reactor. On the other hand, if the parallel reaction to the undesired product D is of higher order than that forming B, a stirred tank would be suggested. These conflicting factors mean that there is no unique answer to the question 'which reactor type will give the highest yield?' The answer will depend on the particular values of the velocity constants and the feed concentration.

Secondly, let us consider the scheme

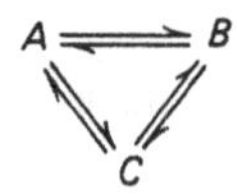

as might be the case in the isomerization of the three xylenes. A number of papers [25–28] have applied matrix or other methods to the treatment of such systems. From this apparently academic mathematical exercise come results of great practical importance, for it is shown how to design experiments to obtain the velocity constants of the scheme in the quickest and most accurate way. Since the determination of kinetic constants for a new industrial process can be a very expensive business indeed, this is a valuable technique.

The evaluation of kinetic constants from experimental reactors holds traps for the unwary [29]. On the other hand, kinetic constants need not always be very precisely known for the adequate design of a reactor. The question of how accurately one needs to know kinetic constants for design is an important one [2, 30].

Lastly it is possible to consider reactions as proceeding by stochastic processes, rather than by deterministic processes. This 'probability' way of looking at reactions has been applied to microbiological reactions [31], which present complicated and coupled schemes.

Example 6.16

Consider the following reactions, which take place in solution:

$A + B \rightarrow C \qquad \text{Rate} = k_1 c_A c_B,$

$C + B \rightarrow D \qquad \text{Rate} = k_2 c_B c_C.$

A batch reaction with solution containing initially 0.1 kmol m^{-3} of A and excess B gave the following analyses:

Time t_1 $\quad c_A = 0.055; \quad c_C = 0.038$ kmol m^{-3};

t_2 $\quad c_A = 0.010; \quad c_C = 0.042.$

What is the ratio of k_2 to k_1? In a batch reaction, what would be the maximum concentration of C? At what conversion of A would this occur?

[*Answers.* 0.52, 0.049 kmol m^{-3}, 74%.]

Example 6.17

Discuss the formation of C by the above reaction,

(*a*) if A is placed in a tank and B added slowly to it,

(*b*) if B is placed in the tank and A added slowly to it.

Example 6.18

A liquid-phase reaction takes place between the reagents A (which is expensive) and B (which is comparatively cheap). The reaction scheme is as follows:

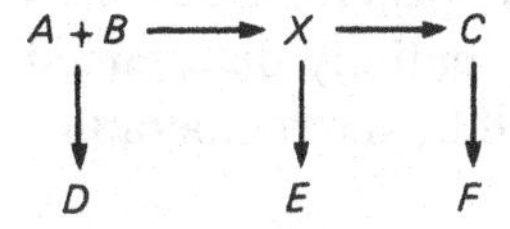

This scheme does not represent either the stoichiometry or the order of the reactions.

(*a*) C is the valuable product; D, E and F are worthless.

(*b*) Water is formed in one or more of the reactions and is known to affect the yield of C adversely.

(*c*) The process is very exothermic, and some of the by-products are tarry and will deposit on surfaces.

Discuss what additional experimental information you would seek to obtain before designing a reactor, and describe how this information would influence your design.

6.9 Instantaneous and overall reaction yields

The concept of a reaction's instantaneous yield, distinct from the overall yield already discussed, is useful for purposes of generalizing some of the points of the previous sections.

Consider the reaction whose stoichiometry is

$$\alpha A + \beta B = \chi X,$$

and let it be supposed that the yield of X is computed relative to reagent A, this being the one which is not in excess. During an infinitesimal period of time $d(A)$ moles of A react to give $d(X)$ moles of X, together with the products of side or consecutive reactions if they occur. The instantaneous yield, ϕ, of X will be defined [**4, 21**] as

$$\phi = \frac{\alpha}{\chi}\frac{d(X)}{d(A)}. \tag{6.20}$$

Here, as in equations (6.1) and (6.2), α/χ is the stoichiometric factor which would make ϕ equal to unity if the wasteful reactions did not occur.

The instantaneous yield is the ratio of the useful rate to the total rate at which reagent is being consumed. If we consider the example in § 6.7, where the rates of the useful and wasteful reactions were taken as being k_1a^2b and k_2ab respectively (and if, for simplicity, we take α/χ as unity) we have

$$\phi = \frac{k_1a^2b}{k_1a^2b + k_2ab} = \frac{k_1a}{k_1a + k_2}. \tag{6.21}$$

The significance of ϕ is that it is the differential yield obtained on each infinitesimal amount of reagent at the moment that it reacts. In general it changes during the course of reaction – or at least whenever concentration changes do not affect the useful and wasteful reactions equally. However, the importance of ϕ is that it is entirely determined by the momentary values of the concentrations and by the temperature, and in this respect it is a function of state.

By contrast the overall yield, Φ or Φ', refers to the amount of product formed at the end of the reaction process (and will be shown to be an integral or summation over the instantaneous yields). As such it is not a function of state but depends on the history of the concentration or temperature changes which the system undergoes; if these variables change in a stepwise manner, as in a C.S.T.R., the overall yield will not be the same as if they change continuously.

Consider first the latter kind of change, as it occurs in a batch or plug-flow reactor. From equation (6.20) it follows that the total amount of X formed at the end of reaction is given by

$$(X) = \frac{\chi}{\alpha}\int_0^{(A)_r} \phi \, d(A). \tag{6.22}$$

Here, as in equation (6.1), $(A)_r$ is the total amount of A that reacts. Hence, by combining equations (6.1) and (6.22), the overall yield obtained in a batchwise or tubular reaction process, Φ'_{BT}, is seen to be the integral over the instantaneous yields:

$$\Phi'_{BT} = \frac{1}{(A)_r}\int_0^{(A)_r} \phi \, d(A). \tag{6.23}$$

Alternatively, using the definition (equation (6.2)) of the overall yield,

$$\Phi_{BT}=\frac{1}{(A)_t}\int_0^{(A)_r}\phi\, d(A). \tag{6.24}$$

Consider now the same reaction carried out in a series of n stirred tanks. Over a given period of time let $\Delta(A)_1$, $\Delta(A)_2$, etc., be the moles of A reacting in the first, second, etc., tanks respectively. If these tanks may be assumed to be perfectly mixed and to be operating at their steady states, the $\Delta(A)_1$ moles reacting in the first tank will give rise to $(\chi/\alpha)\phi_1\Delta(A)_1$ moles of X, where ϕ_1 is the yield characteristic of the steady conditions obtaining in this particular tank. This follows from equation (6.20) as applied to a finite process at constant ϕ. Similarly, with regard to the second tank, etc., and the total amount of useful product in the outflow from the nth tank is therefore

$$(X)=\frac{\chi}{\alpha}\sum_{i=1}^{n}\phi_i\Delta(A)_i. \tag{6.25}$$

Combining this equation with either equation (6.1) or (6.2) we obtain for the overall yield

$$\Phi'_C=\frac{1}{(A)_r}\sum_{i=1}^{n}\phi_i\Delta(A)_i, \tag{6.26}$$

$$\Phi_C=\frac{1}{(A)_t}\sum_{i=1}^{n}\phi_i\Delta(A)_i. \tag{6.27}$$

These expressions, which are the analogues of equations (6.23) and (6.24), show that the overall yield is a summation over the stationary instantaneous yields characteristic of the various tanks, weighted according to the amount of reaction taking place in each.

Since an integral and a summation are not equal in general, the overall yields obtainable by batch or tubular reaction processes are likewise not equal to those obtainable from a chain of stirred tanks. It is seen, however, that the instantaneous yield ϕ provides the logical connection between the different types of process. Whereas, as remarked already, the overall yield depends on the history of the concentration (and temperature) changes, the instantaneous yield is a state function (i.e. a property) of the reacting system and has a unique value for any given values of the concentrations and temperatures.

In the case of reactions whose kinetics are fully understood, the introduction of the instantaneous yield is of no special value. For such reactions the overall yield can always be evaluated and an example was given in § 6.4 above. However, there are many important industrial reactions whose formal kinetics have not been determined and in these instances the use of the instantaneous yield concept can sometimes be of great value. If ϕ can be shown to depend on only one composition variable, one can proceed as shown below. This is particularly valuable since ϕ can often be measured with far less trouble than would be needed for a more formal laboratory study of the kinetics.

We will use for an illustration the nitration of hexamethylenetetramine (hexamine) to form the explosive cyclonite (R.D.X.) [**4**]. Experiments showed that when crystals of hexamine were added to concentrated nitric acid at a suitable temperature the instantaneous yield of cyclonite was almost entirely determined by the momentary concentration of the acid, which becomes progressively more dilute because of the reaction. The reaction appeared to be controlled by the rate of dissolution of the solid hexamine – chemical reaction of the dissolved hexamine seemed almost instantaneous.

It was convenient to define the 'composition variable', ρ, as the weight ratio of the dissolved hexamine to the nitric acid feed. The instantaneous yield was found to be a function of ρ only and it was measured by use of a small C.S.T.R.

For such a system the instantaneous yield characteristic of the conditions prevailing in the tank is equal to the measured overall yield. This follows from equation (6.26), $\Phi'_C = \phi_1$, although it is obvious enough. Yield measurements were obtained from a large number of experimental

Fig. 31. The instantaneous yield in the formation of cyclonite.

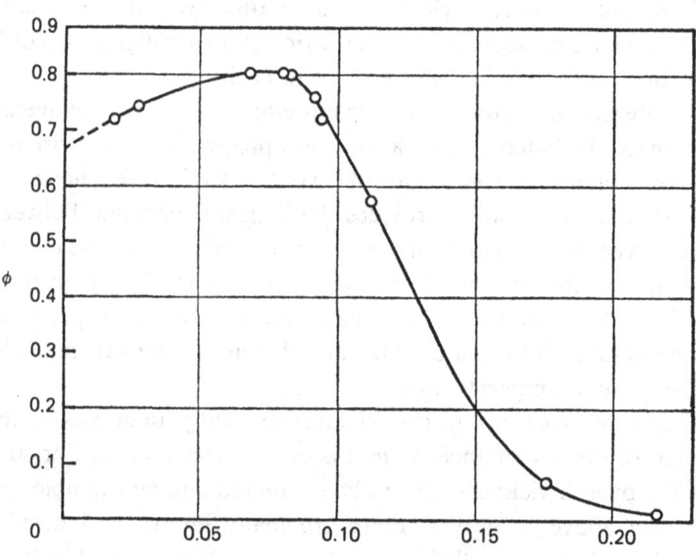

runs, each corresponding to a different choice of the stationary conditions in the tank. The values of ϕ were plotted as a function of ρ, for constant values of the temperature and the initial acid concentration. A curve of this kind referring to a temperature of 25 °C is shown in Fig. 31, and it will be seen that, with increasing ρ, i.e. with increasing dilution of the acid, the value of ϕ first increases a little and then falls very steeply. Evidently there is a certain optimum acid concentration, rather more dilute than the feed acid (98.5% in these experiments), at which the instantaneous yield has its highest value.

In a batch or plug-flow reactor ρ increases continuously up to the end value, ρ_t. It is clear that the appropriate form of equation (6.24) is

$$\Phi_{BT} = \frac{1}{\rho_t}\int_0^{\rho_t} \phi \, d\rho, \tag{6.28}$$

since (A), the quantity of hexamine which has reacted, is proportional to ρ. It will be seen that the integral is equal to the area beneath the curve of Fig. 31 up to any chosen end state ρ_t. That is to say the 'trajectory' of a batch reaction is represented by motion along the curve from left to right, as ρ increases from its initial value of zero up to its final value ρ_t, and instantaneous yields are obtained along the whole trajectory as each increment of hexamine reacts.

Consider now the same process carried out in a chain of n stirred tanks in series. It will be assumed that the quantity $(B)_t$ of acid entering the system in a suitable unit of time, enters it entirely at the first tank. The same assumption will not be made with regard to the hexamine. It will be shown that, because of the very great speed of reaction, there are advantages in distributing the total feed of hexamine over successive vessels. If ρ_1, ρ_2, etc., are successive values of the ratio $(A)/(B)_t$ in the first, second, etc., tanks, we have for the respective quantities of hexamine reacting

$$\left.\begin{aligned} \Delta(A)_1 &= (B)_t(\rho_1 - 0), \\ \Delta(A)_2 &= (B)_t(\rho_2 - \rho_1), \text{ etc.} \end{aligned}\right\} \tag{6.29}$$

It can be assumed that all the hexamine added to each tank is dissolved and reacts in that tank. Substituting equation (6.29) into equation (6.27) we find, for the overall yield as obtained from this type of process,

$$\Phi_C = \frac{1}{\rho_t}\sum_{i=1}^{n} \phi_i(\rho_i - \rho_{i-1}), \tag{6.30}$$

where $\rho_0 = 0$ and $\rho_n = \rho_t$.

It will be seen that the summation in equation (6.30) is equal to the total area of a set of rectangles, equal in number to the number of tanks, each of them touching the ϕ curve at their upper right-hand corners. This is illustrated in Fig. 32, which refers to $n=2$. In the present type of reaction, where ϕ passes through a maximum, the highest overall yield is evidently obtained by using a C.S.T.R. consisting of only a *single* tank provided that the stationary state existing within it can be adjusted to be exactly at the peak (Fig. 33). From a yield standpoint, a batch or tubular reaction process would be distinctly less efficient, as can be seen by comparing the area of the rectangle with the area under the curve.

However, in this example of hexamine nitration the economics of the process depend not only on the yield but also on the amount of excess acid which has to be recovered and recycled. This second factor requires that ρ_t, the terminal value of the ratio of hexamine to acid, must be greater than the value corresponding to the highest value of ϕ, and indeed it must be considerably to the right of the peak. Under these circumstances batch or tubular reaction processes give higher yields than are obtainable from a chain of stirred tanks with the hexamine feed distributed over the chain. The latter may, however, be preferred for other reasons (e.g. ease of cooling).

The optimum values of ρ_1, ρ_2, etc., may be determined by a process of 'maximization of rectangles' [**4, 32**]. For a chosen number n of tanks, and for a chosen terminal ratio $\rho_t(=\rho_n)$, as determined by the economics of acid recovery, the optimal relative quantities of hexamine to be added to the various tanks must be such that the corresponding values of

Fig. 32. A two-stage C.S.T.R. process. (N.B. The rectangles shown are not an optimum choice.) The arrows indicate the trajectory of a batch reaction.

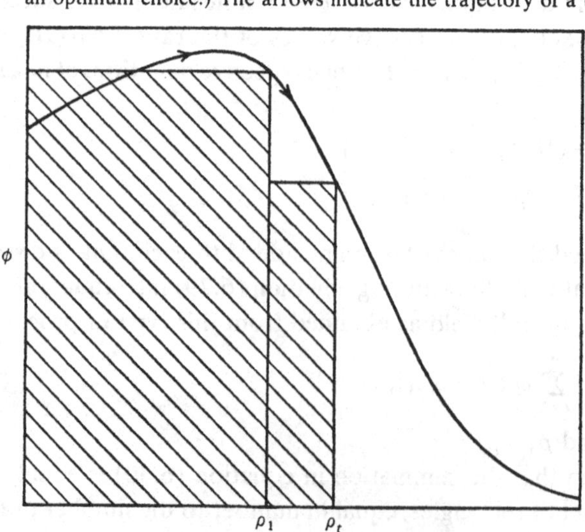

$\rho_1, \rho_2, \ldots, \rho_{n-1}$, make the area of the set of n inscribed rectangles equal to a maximum. Given the shape of the ϕ curve from experiment, it is a relatively simple procedure to find the position of the $n-1$ ordinates which satisfy this condition.

This example of hexamine nitration has been discussed in some detail as an instance of the utility of ϕ measurements as applied to a certain class of reactions. In other examples it might occur that the curve, instead of passing through a maximum, diminishes steadily with increased progress of reaction. In this situation a batch or tubular reactor would inevitably give a better yield than any possible combination of stirred tanks. If the latter had to be used for quite different reasons, it would clearly be desirable to use as many tanks in series as could reasonably be allowed. Conversely, if the ϕ curve were found to rise steadily with increasing progress of the reaction, a single-tank C.S.T.R. should be chosen if yield considerations were dominant. No doubt it might be necessary in practice to use two or more tanks in series, in order to reduce the bypassing loss; however, the first tank should be chosen much larger than the others in order that the reacting system should dilute itself as much as possible at an early stage.

6.10 Combination of C.S.T.R. and tubular reactor

It is clear from the previous section that from the standpoint of yield the best arrangement in a case such as is shown in Fig. 31 would

Fig. 33. A single-stage C.S.T.R. process with ρ adjusted to its optimum value.

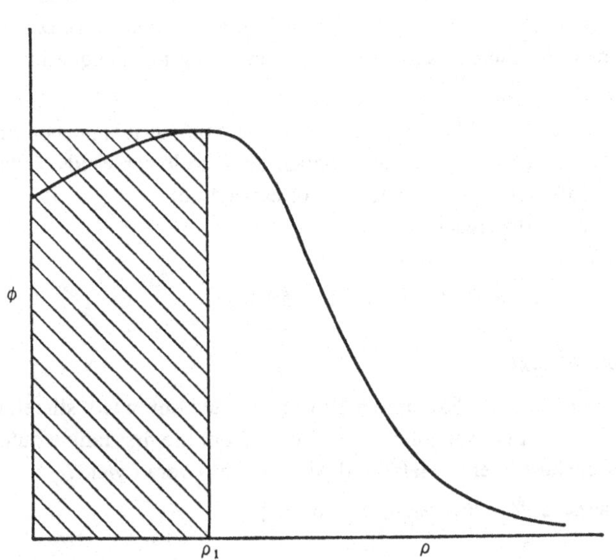

be a stirred tank operating as in Fig. 33, followed by a tubular reactor in which ρ moves from ρ_1 to ρ_f. This arrangement of a stirred tank followed by a tubular reactor was also found to be the best in § 4.4 for an autocatalytic reaction. In the latter case the criterion for 'best' was that of minimum reactor total volume, i.e. a *rate* criterion. We are, in this chapter, concerned with the analogous *yield* criterion.

Another situation in which the yield can be increased by a combination of reactors occurs in the system of three parallel reactions $A \to B$, $A \to C$ and $A \to D$, where the orders of the reactions are in the sequence

$$A \to B < A \to C < A \to D$$

and C is the desired product. It will be evident, from what was said earlier, that as regards competition between B and C as products, the tubular reactor would be preferable, whereas as regards competition between C and D the single-tank C.S.T.R. would be the best choice. Moreover, because of the different influence of the concentration of A, the relative significance of the two forms of competition changes as reaction proceeds; thus wasteful formation of D tends to be more important during the early stages of reaction and wasteful formation of B becomes more important during the later stages. For this reason an improved yield can be obtained by using a stirred tank followed by a tubular reactor [**8**]. Other similar situations have also been discussed [**33, 34**].

Example 6.19

A product, B, is to be produced from a reagent, A, which also reacts to give worthless by-products. The instantaneous yield of B, ϕ, is found to depend on the conversion of A, x, according to the equation $\phi = 0.6 + 2x - 5x^2$.

The reaction will be terminated when $\phi = 0.5$, as it is uneconomic to continue to higher conversions, which leads to lower yields. Calculate the overall yield if the reaction were carried out

(a) in a batch reactor,

(b) in a single stirred tank.

[*Answers.* (a) $\Phi' = 71.5\%$, (b) $\Phi' = 50\%$.]

Example 6.20

If the reaction in Example 6.19 were carried out in two stirred tanks in series, what conversion in the effluent from the first tank would lead to the highest overall yield, and what is that overall yield?

[*Answers.* 29% conversion, $\Phi' = 67\%$.]

***Example* 6.21**

If the effluent from the plant is to correspond to 50% conversion, what is the highest overall yield which could be obtained, and what arrangement would produce it?

[*Answer.* $\Phi' = 74.4\%$.]

6.11 Removal of product. Recycle

In this chapter we have mainly considered irreversible reactions. Many reactions do not, however, go to completion, but rather to an equilibrium state, which may be far from complete conversion. For an example, let us consider the reversible reaction

$$A + B \rightleftharpoons C + D.$$

If A is a more expensive raw material than B, it could well be advantageous to operate with a considerable excess of B. The conversion of A would then be more complete when equilibrium is approached. The product stream would consist of C and D, with a small amount of A, and a considerable quantity of excess B. In the purification of the products C and D, A and B would be separated and recycled to the reactor.

In other cases it may be possible to remove C (and/or D) from the reaction mixture, for example by distilling it off, if it is markedly more volatile than the other components. This is done in the manufacture of nylon 66, where the polymerization of adipic acid and hexamethylene diamine would not go to completion. However, one of the products of the reaction is water, which can be readily distilled from the mixture of reagents ('nylon salt') and polymer.

Another type of behaviour is when the desired reaction, $A \rightarrow B$, is succeeded by an undesired reaction, $B \rightarrow C$, and the kinetics of these reactions demand a low pass conversion if the yield of B is to be acceptable, i.e. to maintain a high 'material efficiency' of conversion of A to B. The nylon process provides an example of this, too. The oxidation of cyclohexane to 'KA' (a mixture of the *k*etone and *a*lcohol of cyclohexane) is a first step in the manufacture of adipic acid. This reaction is run at a pass conversion of less than 10%, since higher attempted conversions of the cyclohexane lead to unacceptable losses of the KA by further oxidation. It is thus essential to separate large quantities of unreacted cyclohexane from the reactor effluent, and recycle this to the reactor inlet.

The use of recycle streams (with provision of purge streams if necessary) is so widespread in industry that calculations on such processes are among the first taught to student chemical engineers. The *design* of such systems involves a blend of chemical reaction engineering and economic optimization to produce the most cost-effective process. For a further treatment of recycle systems see [**35–37**].

Example 6.22

An isothermal reaction $A \rightleftharpoons B$ is carried out in aqueous solution. The reaction is first order in both directions, with velocity constant $k_{A \to B} = 0.4\ \text{h}^{-1}$, and the equilibrium constant is 4.0. The feed to the plant contains $100\ \text{kg m}^{-3}$ of A and $12\ \text{m}^3\ \text{h}^{-1}$ of this feed are to be treated. The reactor is a stirred tank of volume $60\ \text{m}^3$, and the effluent passes to a separator, in which B is recovered. A fraction, y, of the unreacted A is recycled to the reactor as a solution containing $100\ \text{kg m}^{-3}$ of A; the remaining A and water are rejected.

The product B is worth £0.20 per kg. The operating costs are £5.0 per m^3 of solution entering the separator. What value of y maximizes the operational profit of the plant, and what fraction of the A fed to the plant is converted at this optimum?

[*Answers.* $y = 0.89$, 88% conversion.]

Example 6.23

A substance A is isomerized by a first-order reaction to B in a gas-phase C.S.T.R. at 1 bar. It also undergoes the second-order disproportionation reaction $2A \to C + D$, giving worthless gaseous products. The effluent from the C.S.T.R. is separated into its components, the unused A being recycled as a pure gas at 1 bar. The conversion per pass of A is 60%, and the material efficiency of the process, to useful product B, is 80%. If the same quantity of A is to be supplied to the plant at the same temperature, but the pressure is raised to 2 bar, what will be the new conversion per pass and material efficiency?

[*Answers.* 80%, 80% (as before).]

Example 6.24

Compound A is decomposed in a tubular reactor according to the following first-order isothermal irreversible reactions:

$$A \xrightarrow{k_1} X \xrightarrow{k_2} Z, \qquad X \xrightarrow{k_3} Y$$

The feed rate of A is $16\ \text{m}^3\ \text{min}^{-1}$. If unconverted feed A and product Y have negligible cash value and product X has twice the cash value of product Z, determine the optimum volume of the reactor and the yields of X and Z for this condition.

Assume (*a*) volume changes are negligible, (*b*) plug-flow conditions obtain, and (*c*) $2k_1 = 3k_2 = k_3 = 6\ \text{min}^{-1}$.

[*Answers.* $4\ \text{m}^3$, 20.4%, 7.7%.]

Example 6.25

A reaction proceeds in a batch reactor according to the following first-order steps

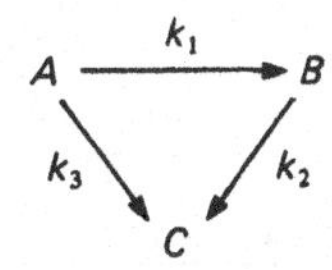

where B is the desired product and C is a waste product. The initial concentrations of B and C are zero. Show that the concentration of B is a maximum at time

$$t_m = \frac{\ln[k_2/(k_1+k_3)]}{k_2 - k_1 - k_3}$$

provided $k_2 \neq k_1 + k_3$. Find the expression for t_m when $k_2 = k_1 + k_3$.

For $k_1 = 10\ \text{h}^{-1}$, $k_2 = 5\ \text{h}^{-1}$ and $k_3 = 2\ \text{h}^{-1}$ calculate the maximum conversion of A to B.

[*Answers.* $t_m = 1/k_2$, 0.446.]

Symbols

a, b	Concentration of A, B, kmol m^{-3}.
c	Concentration, kmol m^{-3}.
k	Velocity constant, units depend on order of reaction.
r	Number of monomer molecules in polymer.
s	Ratio of velocity constants, see equation (6.11).
t	Time, s.
v	Volume flow rate, $\text{m}^3\ \text{s}^{-1}$.
V	Volume of reactor, m^3.
α, β, χ	Stoichiometric coefficients.
ρ	Composition variable – see § 6.9.
Φ, Φ'	Overall yields, defined by equations (6.1), (6.2).
ϕ	Instantaneous yield, see equation (6.20).

References

1. MacMullin, R. B., *Chem. Engng Progr.*, 1948, **44**, 183.
2. Tichacek, L. J., *A.I.Ch.E. Journal*, 1963, **9**, 394.
3. Denbigh, K. G., *Inst. of Chem. Eng.*, Symposium on Scaling-Up, 1957.
4. Denbigh, K. G., *Chem. Engng Sci.*, 1961, **14**, 25.
5. Kramers, H. and Westerterp, K. R., *Elements of Chemical Reactor Design and Operation* (Netherlands University Press, 1963), pp. 80 ff.
6. Adler, J. and Vortmeyer, D., *Chem. Engng Sci.*, 1964, **19**, 413.
7. Horn, F. J. M. and Parish, T. D., *Chem. Engng Sci.*, 1967, **22**, 1549.
8. Trambouze, P. J. and Piret, E. L., *A.I.Ch.E. Journal*, 1959, **5**, 384.
9. Wei, J., *Canad. J. Chem. Engng*, 1966, **44**, 31.
10. Chermin, H. A. G. and van Krevelen, D. W., *Chem. Engng Sci.*, 1961, **14**, 58.
11. Shoenemann, K., *Chem. Engng Sci.*, 1961, **14**, 39.
12. Rys, P., *Angew. Chem.*, 1977, **89**, 847.
13. Denbigh, K. G., *Trans. Faraday Soc.*, 1947, **43**, 648.
14. Denbigh, K. G., *J. Appl. Chem.*, 1951, **1**, 227.
15. Horikx, M. M. and Hermans, J. J., *J. Polym. Sci.*, 1953, **11**, 325.
16. Villermaux, J. and David, R., *Chem. Engng Comm.*, 1983, **21**, 105.
17. Bransom, S. H., Dunning, W. J. and Millard, B., *Disc. Faraday Soc.*, No. 5, 1949, 83.
18. Mullin, J. W., *Crystallization*, 2nd Edn. (Butterworths, London, 1972).
19. Mattern, R. V., Bilous, O. and Piret, E. L., *A.I.Ch.E. Journal*, 1957, **3**, 497.
20. Kunii, D. and Levenspiel, O., *Fluidization Engineering* (John Wiley, New York, 1969), Chapter 11.
21. Denbigh, K. G., *Trans. Faraday Soc.*, 1944, **40**, 352.
22. van de Vusse, J. G. and Voelter, H., *Chem. Engng Sci.*, 1961, **14**, 90.
23. Grutter, W. F. and Messikommer, B. H., *Chem. Engng Sci.*, 1961, **14**, 321.
24. van de Vusse, J. G., *Chem. Engng Sci.*, 1964, **19**, 994.
25. Wei, J. and Prater, C. D., *Adv. Catalysis*, 1962, **13**, 203.
26. Prater, C. D., Sylvestri, A. J. and Wei, J., *Chem. Engng Sci.*, 1967, **22**, 1587.
27. Faith, L. E. and Vermeulen, T., *A.I.Ch.E. Journal*, 1967, **13**, 936.
28. Himmelblau, D. M., Jones, C. R. and Bischoff, K. B., *Ind. Eng. Chem. Fund.*, 1967, **6**, 539.
29. Wei, J., *Ind. Eng. Chem.*, Sept., 1966, 38.
30. Ray, W. H. and Aris, R., *Ind. Eng. Chem. Fund.*, 1966, **5**, 478.
31. Frederickson, A. G., *Chem. Engng Sci.*, 1966, **21**, 687.
32. Aris, R., *Chem. Engng Sci.*, 1960, **13**, 75.
33. Cholette, A. and Blanchet, J., *Canad. J. Chem. Eng.*, 1961, **39**, 192.
34. Dialer, K., Horn, F. and Kuchler, L., *Chemische Reaktionstechnik*, p. 295, in Winnacker-Kuchler, *Chemische Technologie*, Band I (Carl Hauser Verlag, Munich, 1958).
35. Nagiev, M. F., *The Theory of Recycle Processes in Chemical Engineering* (Pergamon Press, Oxford, 1964).
36. Gillespie, B. and Carberry, J. J., *Ind. Eng. Chem. Fund.*, 1966, **5**, 164.
37. Schmeal, W. R. and Amundson, N. R., *A.I.Ch.E. Journal*, 1966, **12**, 1202.

7

Packed-bed catalytic reactors: mass- and heat-transfer effects

7.1 Heterogeneous catalysis

Packed-bed catalytic reactors are of the greatest economic importance, and constitute the backbone of the chemical industry. Ammonia synthesis, sulphuric acid production and petroleum refining are three processes of huge tonnage which use packed-bed catalytic reactors. Many such reactors of lower throughput, but producing materials of high value, are to be found in, for example, the petrochemical industry.

The presence of the catalyst packing is, of course, vital to the reaction, which occurs 'on the surface of the catalyst'. The observed reaction rate depends on the nature of the surface, on how much of it is present in the reactor, and on the ability of the molecules to get to and away from it.

The nature of the surface, whether it can act as a catalyst at all, is not a subject for us to discuss. This is a field of physical chemistry in which much work has been, and is being, done. It must be said that so far the ability of such work to *predict* catalytic activity is limited, though it is possible to discover general trends. Catalysts are discovered by trial and error, but it is now possible to have very shrewd ideas where to look for success. Catalyst surfaces are very easily 'poisoned' by small amounts of deleterious substances. Equally, their activity can often be improved by relatively mild measures; e.g. the activity of silica–alumina catalysts can be altered by treating the catalyst with steam.

The most fundamental laboratory studies on catalysts are carried out on plane surfaces, e.g. of platinum, mercury, etc., and the measured reaction rates are then most naturally reported per unit area of the catalyst. Industrial catalysis, on the other hand, is usually carried out on porous material* since this results in a much larger amount of active area per unit of reactor volume. Since the effective internal area of the catalyst pellet (as distinct from its external surface area) may not always be exactly known, it is usually most convenient to express the reaction rate as being per unit mass of the catalyst. When so expressed, the rate will clearly depend on the porosity of the pellet, and therefore the same catalytic substance, when made up as a pellet from powder in two different ways, may result in two quite different rates. The internal surface area can often be measured by the gas adsorption techniques of surface chemistry. The effectiveness of this apparently available area depends on other factors. § 7.5 discusses this more fully.

As well as depending on catalyst porosity, the reaction rate per unit mass will be some function of the reagent concentrations and of the temperature, but this function may not be as simple as in the case of uncatalysed reactions. Before catalysis can take place, the reagents have to diffuse through the pores and it can occur that either reaction or diffusion is the rate-limiting process, or that both of them have an almost equal effect. If reaction is rate-limiting, as tends to occur in a lower temperature range, the effects of concentration and temperature will be those which are typical of chemical reaction. Conversely, if diffusion is rate-limiting, as tends to occur in a higher temperature range, the effects of concentration and temperature will be those which are characteristic of diffusion. In the transitional region, where both reaction and diffusion affect the overall rate, the influence of temperature and concentration is often rather complex.

The reason why chemical reaction on the surface tends to limit the reaction rate in a lower temperature range, while diffusion limits in a higher range, is that chemical reaction has usually a very much greater temperature coefficient than diffusion. The overall rate thus increases rapidly with increasing temperature at low temperatures as the chemical reaction on the surface speeds up, but eventually, at higher temperatures, the diffusional processes cannot supply the surface reaction with the desired reactants (or perhaps remove the products) fast enough. The overall rate subsequently increases less rapidly with increasing temperature.

Two further points should be mentioned here. First, the effectiveness of a catalyst lies in its ability to provide a reaction path of reduced activation energy (even at a cost of much lower 'collision number').

* The formulation of catalyst pellets will be discussed in § 7.7.

Secondly, the interaction between diffusion and reaction means that laboratory kinetic data need to be very carefully interpreted [**1**].

7.2 Mass transfer between packing and fluid

It is customary to consider the product from a packed-bed reactor as being the result of the following sequence of events:

(i) Mass transfer of reagent from the fluid phase to the external surface of the catalyst packing.
(ii) Diffusion into the pores of the catalyst particles.
(iii) Adsorption onto the catalyst surface.
(iv) Reaction on the surface.
(v) Desorption of product from the catalyst surface.
(vi) Diffusion out from the pores of the catalyst particles.
(vii) Mass transfer of product into the fluid phase.

We have briefly touched on (iii), (iv) and (v) in the previous section. Steps (i) and (vii) will now be discussed, whilst consideration of (ii) and (vi) will be deferred to § 7.3. Steps (i) and (vii) are, of course, significant only when the fluid is a mixture – there is no diffusional resistance to mass transfer from a *pure* fluid to a solid immersed in it. We shall for the moment assume that the catalyst and surrounding fluid are all at the same, constant, temperature.

A convenient model for such mass transfer is the 'stagnant-film' model, in which it is proposed that close to the pellet surface is a stagnant film of fluid, through which reagent has to diffuse in order to react at the pellet. The fluid beyond this stagnant film is assumed to be maintained at uniform composition by the effects of stirring or of fluid turbulence. The fluid mechanics also determine the (effective) thickness of the stagnant film; the diffusion properties of the materials determine the transport through it.

In the steady state (which is almost always assumed in these studies) the rate of transport of reagent to the surface must equal its reaction rate on the surface. If, for simplicity, we consider a single reagent, we will have

$$k_M(c_b - c_s) = kc_s^n, \qquad (7.1)$$

in which both sides of the equation are based on unit *area* of external pellet surface, the reaction is of nth order in the concentration at the surface, c_s, and k_M is the mass-transfer coefficient from the bulk fluid, of concentration c_b, to the surface. If the reagent is at low concentration in some other, inert, substance, then $k_M = D/\delta$, where D is the appropriate (binary) diffusion coefficient, and δ the apparent stagnant-film thickness.

An equation similar to equation (7.1) can be written for diffusion of product *away* from the surface (step (vii)). For reactions which, at equilibrium, are virtually complete the effects of product diffusion are not usually serious; it is more likely that product accumulation at the catalytic surface (step (v)) will prove troublesome.

It is clear from equation (7.1) that c_s is less than c_b. We may define an effectiveness factor, η, by

$$\eta = \frac{\text{observed rate of reaction}}{\text{rate obtainable if } c_s \text{ were equal to } c_b}. \tag{7.2}$$

We may also define the Damköhler number, *Da*, as the ratio of the maximum possible reaction rate (kc_b^n), to the maximum possible transport rate ($k_M c_b$).

***Example* 7.1**

Show that if $n = 1$, $\eta = (1 + Da)^{-1}$.

***Example* 7.2**

Show that if $n = 2$,

$$\eta = \frac{1}{2Da}((1+4Da)^{1/2} - 1)^2.$$

The effective film thickness, δ, or the mass-transfer coëfficient, k_M, have to be empirically determined, and a correlation for such mass transfer which has often been proposed has the form

$$Sh = k_M d/D = A Re^{1/2} Sc^{1/3}, \tag{7.3}$$

in which A is a dimensionless constant and d is the (equivalent) diameter of a packing piece. *Sc* is the Schmidt number; the Reynolds number, *Re*, is based on d and on the *superficial* velocity (i.e. the velocity the fluid *would* have if there were no packing in the reactor). Values of A are of order unity: experiments on packed beds of spheres [**2**] gave $A = 1.9$, theories based on mass transfer through a developing boundary layer gave $A = 1.5\epsilon^{-1/2}$ [**3**] or $0.974\epsilon^{-1/2}$ [**4**], where ϵ is the voidage of the bed. The range of voidage in packed beds is not very large, and experimental results on packed and aggregatively fluidized beds, which have a wider range of ϵ (see later), gave $A = 0.81\epsilon^{-1}$ [**5**]. If the voidage is about 0.4, the difference between these values of A is small and can for our purposes be neglected.

The experimental results for *Sh* considered in [3] clustered within ±20% or so of the theoretical line, except for one set, at high *Sc*, which were rather higher. The range of *Re* covered was from about 0.5 to 1000.

The form of equation (7.3) is similar, not surprisingly, to correlations for gas-absorption in irrigated packed towers where 'the gas-film controls', though the latter usually involve powers of *Re* rather greater than $\frac{1}{2}$. For reactor design, equation (7.3) can be used to estimate the maximum rate at which reagent can diffuse to the outer surface of the catalyst particles, and this represents an upper bound to the reaction rate.

Example 7.3

A binary gas mixture is flowing with superficial velocity 0.1 m s^{-1} through a bed of 5 mm diameter catalyst spheres. The gas has density 1 kg m^{-3} and viscosity 3×10^{-5} N s m^{-2}, and the diffusion coefficient is 4×10^{-5} m^2 s^{-1}. Use equation (7.3) to estimate the value of Sh and k_M for mass transfer between the gas stream and the packing.

Solution

The Schmidt number for this mixture is given by

$Sc=\mu/\rho D=3\times10^{-5}/1\times4\times10^{-5}=0.75.$

The Reynolds number

$$Re=\rho ud/\mu=\frac{1\times0.1\times5\times10^{-3}}{3\times10^{-5}}=16.7.$$

If we use equation (7.3), with $A=1.9$, we find

$Sh=1.9\times16.7^{1/2}\times0.75^{1/3}=7.06.$

The mass-transfer coefficient k_M is given by

$k_M=Sh\cdot D/d=7.06\times4\times10^{-5}/5\times10^{-3}=5.65\times10^{-2}\ \mathrm{m\,s^{-1}}.$

Example 7.4

A binary *liquid* mixture is flowing with superficial velocity 0.1 m s^{-1} through a bed of 5 mm diameter catalyst spheres. The liquid has density 1000 kg m^{-3} and viscosity 10^{-3} N s m^{-2}, and the diffusion coefficient is 8×10^{-10} m^2 s^{-1}. Check that these are reasonable values of the properties, and estimate Sh and k_M. Compare the answers with those from Example 7.3.

[*Answers.* 460, 7.4×10^{-5} m s^{-1}.]

Example 7.5

In Example 7.4 the catalyst bed has a volume of 2 m^3 and a voidage of 0.4. If the concentration of reagent in the liquid stream is 1 kmol m^{-3}, what is the *maximum* rate of reaction of the reagent in the reactor?

Compare this answer with that which you would calculate from Example 7.3 for a similar catalyst bed through which a *gaseous* mixture containing reagent at concentration 10^{-2} kmol m^{-3} is passed. (Check what *partial pressure* this might correspond to.)

[*Answers.* 0.106 kmol s^{-1} and 0.81 kmol s^{-1}.]

As will have been seen from working out the examples, the much lower diffusion coefficients of liquids, as compared with gases, lead more readily to diffusion limitations in reaction rate. Again, high values of *Re* are more commonly encountered with gas flow through packed beds than with liquid flow. Equation (7.3) shows that this also will lead to higher mass-transfer rates with gases, though it is true that *concentration* gradients can more readily be made large with liquids than with gases (except when high pressures are adopted in the latter case). If the reaction rate on the surface is very high indeed, even a gas reaction can become controlled by mass transfer to the surface of the catalyst. The oxidation of ammonia over a platinum gauze is an important example, being the first step in the production of nitric acid from ammonia.

If the chemical reaction is 'slow', it is usually necessary to employ a porous catalyst if there is to be enough active surface area to obtain a worthwhile rate in an acceptable bed volume. In that case, when the total surface area is perhaps several orders of magnitude greater than the pellet external surface area, diffusional phenomena within the pellet will be of much greater importance than mass transfer to the pellet external surface. Thus the equations of this section are not commonly of much practical relevance, but they do serve as a useful introduction to the subject matter of the next section.

7.3 Diffusion within catalyst particles

At the beginning of the previous section we considered the seven steps involved in reaction within a packed-bed catalytic reactor. Steps (ii) and (vi) involve diffusion of reagents and products within catalyst pores. Highly active catalyst pellets, obtained by compressing finely divided powder, usually possess a large amount of internal surface. Such pellets are porous, and the area available 'within' a pellet is vastly greater than its apparent external area – perhaps as much as a million times greater. It is, of course, necessary for reagent to succeed in reaching this internal surface, and this will involve diffusion within the pores.

The magnitude of the diffusion coefficient to be used in the subsequent equations of § 7.4 requires careful consideration. If the pores are large and/or the molecular density of the fluid within them is high, the diffusion coefficient will be the same as in the bulk fluid, with two important modifications. The first one is because the voidage of the pellet has to be allowed for when considering the flux across any elementary cross-section. The second is due to the pores being neither straight nor of uniform cross-section, and therefore a 'tortuosity factor' has to be introduced. An effective diffusion coefficient, D_{eff}, can be defined by

$$D_{eff} = D\epsilon/\tau, \tag{7.4}$$

where ϵ is the voidage and τ is the tortuosity factor. This latter factor cannot really be calculated, though if D_{eff} is *measured* the derived value of τ should be found to lie between 1 and 10 if the above picture concerning tortuosity is reasonably applicable to the particular catalyst.

If the pores are small and/or the molecular density of the fluid is low, then the molecules in a pore will collide with the walls of the pore far more frequently than with each other. Under these conditions Knudsen diffusion takes place, and the effective diffusion coefficient is given by

$$D_{eff} = 100r\sqrt{(T/M)}, \tag{7.5}$$

in which M is the molecular weight, D_{eff} is in $m^2\,s^{-1}$, r, the (equivalent) pore radius, is in m, and T is the absolute temperature.

If we consider a fluid in the region of intermediate molecular density, the effective diffusion coefficient is given by neither equation (7.4) nor (7.5), but by a more complicated expression [**6**]. Also the possibility of mass transfer by diffusion in an adsorbed surface layer cannot be ruled out. For this to be significant, surface adsorption must be considerable, and the adsorbed molecules must be mobile. High-temperature (chemisorption) will not give rise to this, but where there is physical adsorption, as occurs at lower temperatures, surface diffusion may be appreciable [**7**].

A monograph by Satterfield and Sherwood [**8**] gives a good summary of the methods available for estimating D_{eff} in a catalyst particle. Since that date, Satterfield has published two books [**9, 10**] on the subject of heterogeneous catalysis. The problem of estimating D_{eff} is by no means cleared up, as papers at the Chemical Reaction Engineering Symposium in 1980 [**11**], showed.

7.4 Catalyst effectiveness

Since reagent has to diffuse through the outer pores of a catalyst particle to react on the surface of the inner pores, and diffusion requires a concentration gradient, this means that the concentration of reagent in the fluid phase will be less in the centre of a catalyst particle than at its outer surface. Hence the rate of any reaction will in general be less at the centre than at the outside surface.

The *effectiveness factor*, η, as already defined in § 7.2, is the ratio of the observed rate to that which would be obtained if the whole of the internal surface of the pellet were available to the reagents at the same concentrations as they have at the external surface.

It has been shown by Thiele [**12**] and Zeldovitch [**13**], among others, that η is a function of the Thiele modulus ϕ, where

$$\phi = d(kc^{m-1}\sigma/D_{\text{eff}})^{1/2} \tag{7.6}$$

Here d is a measure of the pellet size, k is the reaction velocity constant per unit area, c is the reagent concentration, m the order of reaction, and σ is the surface area of the catalyst per unit volume. D_{eff} has been discussed in the last section; we shall from now on drop the subscript, and refer to D.

For a *first-order* reaction ϕ becomes

$$\phi = d(k\sigma/D)^{1/2}, \tag{7.7}$$

which is independent of concentration.

It is instructive to show the significance and usefulness of the Thiele modulus. We consider an irreversible, first-order reaction in a uniform spherical pellet, and assume that isothermal conditions obtain. (These limitations will be reconsidered later.)

Consider Fig. 34, which shows a cross-section of a sphere in which reaction and diffusion of a reagent are taking place. Let the concentration of the reagent be c, and let it be assumed that the situation is a steady-state one, as could be the case in a continuous tubular reactor.

The mass balance equation for the reagent in the spherical shell of radius r to $r+dr$ is

$$-4\pi r^2 \cdot D\left(\frac{\partial c}{\partial r}\right)_r + 4\pi(r+dr)^2 \cdot D\left(\frac{\partial c}{\partial r}\right)_{r+dr}$$

Inner area · Flux F_1 Outer area · Flux F_2

$$= 4\pi r^2\, dr \cdot k\sigma c^m; \tag{7.8}$$

Volume · Reaction rate per unit volume

Fig. 34. Model for diffusion and reaction in a sphere.

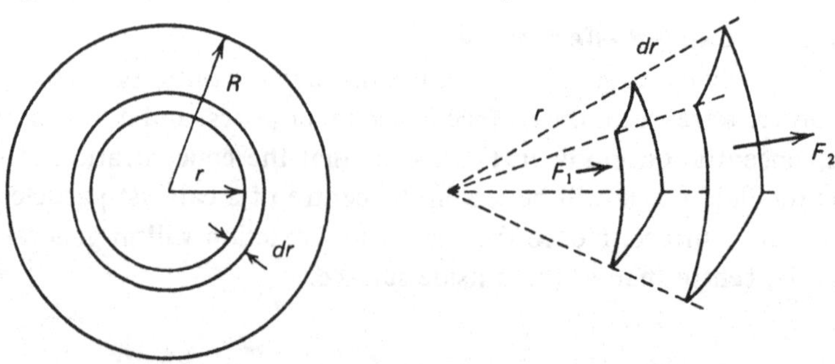

there being no accumulation term since we have a steady state. Also

$$\left(\frac{\partial c}{\partial r}\right)_{r+dr}=\left(\frac{\partial c}{\partial r}\right)_r+\left(\frac{\partial^2 c}{\partial r^2}\right)_r\cdot dr.$$

Substituting in equation (7.8) and neglecting second-order small terms we obtain

$$\frac{\partial^2 c}{\partial r^2}+\frac{2}{r}\frac{\partial c}{\partial r}=\frac{k\sigma c^m}{D}. \tag{7.9}$$

Equation (7.9) can be written as

$$\frac{\partial^2 c}{\partial r^2}+\frac{2}{r}\frac{\partial c}{\partial r}=\frac{\phi^2}{R^2}c, \tag{7.10}$$

in which it can be seen that $\phi=R(k\sigma c^{m-1}/D)^{1/2}$ and is a Thiele modulus – see equation (7.6). For a first-order reaction $m=1$ and ϕ can be taken to be constant, but this will not be so for any other order of reaction.

The solution of this equation is straightforward for the first-order reaction and for the simple boundary conditions:

$$r=0,\qquad \partial c/\partial r=0 \quad\text{and}\quad r=R,\qquad c=c_s,$$

where c_s is the concentration at the exterior surface of the pellet and is equal to the bulk fluid concentration if the mass-transfer resistance of § 7.2 is negligible.

Example 7.6

Show that the solution of equation (7.10) subject to the above boundary conditions is

$$c=\frac{c_s R \sinh(\phi r/R)}{r\sinh\phi}. \tag{7.11}$$

Given equation (7.11) we can find the total amount of reaction going on in the pellet by noting that this quantity equals the diffusive flux entering the outer surface of the pellet.

$$\text{Reaction rate per pellet}=4\pi R^2 D\left(\frac{\partial c}{\partial r}\right)_R. \tag{7.12}$$

We can substitute from equation (7.11) in equation (7.12) to obtain the reaction rate, and can then compare the answer with the rate which would be expected if the concentration inside the pellet were everywhere c_s. This latter quantity is clearly $\frac{4}{3}\pi R^3\cdot k\sigma c_s$. The ratio of the actual to this maximum rate is the effectiveness factor η.

Example 7.7

Show that the effectiveness factor, η, for a first-order reaction in an isothermal, spherical catalyst pellet is given by

$$\eta=\frac{3}{\phi}\left(\frac{1}{\tanh\phi}-\frac{1}{\phi}\right). \tag{7.13}$$

Example 7.8

Show that when ϕ is small, $\eta \to 1-\phi^2/15$ for the case considered in Example 7.7.

7.5 Effectiveness factors

The effectiveness factor, η, for a first-order irreversible reaction in an isothermal, spherical catalyst pellet is given by equation (7.13). For large values of ϕ, which is when the reaction is 'diffusion-controlled', we see from equation (7.13) that $\eta \to 3/\phi$, and the effectiveness factor varies as $D^{1/2}/R$.

We have now to consider what ϕ will be:

(*a*) if the particle is not a sphere;

(*b*) if the reaction is not first-order and reversible;

(*c*) if the pellet is not isothermal but, say, hotter in the middle, due to the reaction.

If the pellet is not spherical (and it usually is not) the above treatment is not applicable in detail. Certain simple geometries can be analytically treated, including the straight pore, the planar slab, or the cylinder with sealed ends. In fact Aris [**14**] has shown that the results for all geometries are unlikely to be significantly different (remember that D is unlikely to be accurately known) if the characteristic length, d, of equation (7.6) is chosen to be the ratio of the total volume to the external surface area exposed to the fluid. For a sphere, we therefore take $d = R/3$, while for a slab d = half the slab thickness, and for a cylinder with sealed ends d = half the cylinder radius. With these values of d in equation (7.6) then in all three cases

$$\eta \to 1/\phi \quad \text{for large } \phi. \tag{7.14}$$

Fig. 35. 'Common plot' of effectiveness factor η versus Thiele modulus ϕ for particles of different shapes.

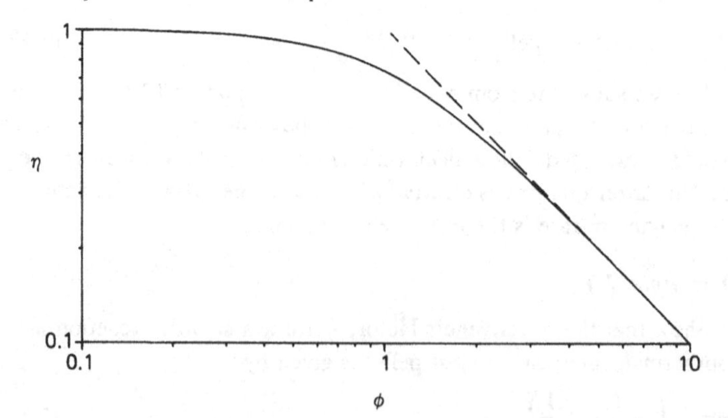

Fig. 35 shows a 'common plot' of η against ϕ, to be used for any geometry if d is chosen as above. For values of ϕ between about 0.4 and 2, the plots of η for different geometries are different, but only by a few %. For other values of ϕ, the 'common plot' is an accurate picture, for all geometries.

For reactions of other orders, equation (7.10) becomes non-linear – since ϕ depends on c (see equation (7.6)). For simple orders and geometries, solution is possible, but Aris [**15**] has shown that if the Thiele modulus is defined somewhat differently, then – whatever the kinetics – the asymptotic value of η obeys equation (7.14). Thus Fig. 35 is a useful plot for any kinetics or geometry, provided ϕ is defined appropriately as in [**14**, **15**].

A comprehensive treatment of diffusion in permeable catalysts has been given by Aris [**16**], to which the advanced student is directed.

If the pellet is not isothermal, startling deviations from these results can be obtained – as will now be described in the following section.

7.6 Thermal effects in catalyst particles

The effect of temperature on the rate of chemical reaction is often so large that if conditions are not maintained isothermal, the temperature distribution is the dominant factor in determining reactor performance. We have considered this in Chapter 2, on chemical kinetics, and in Chapter 3, on tubular reactors, and we shall consider it again in Chapter 9. We shall here treat only the effect of the temperature differences associated with heat transfer within and from a catalyst pellet.

Firstly we will deal with heat transfer within a catalyst pellet. The most important cases occur with exothermic reactions, where heat is generated within the pellet, is conducted to the outer pellet surface and is transferred to the surrounding fluid. The magnitude of the temperature difference between the interior and surface of the particle can be estimated quite simply, as has been shown by Prater [**17**]. Let c be the concentration of reagent at any point within the porous body of the particle, and let $\dot{n}$ be the rate of consumption of reagent by reaction at this point, per unit of catalyst volume. Then from a mass balance

$$D\nabla^2 c \equiv D\left(\frac{\partial^2 c}{\partial x^2}+\frac{\partial^2 c}{\partial y^2}+\frac{\partial^2 c}{\partial z^2}\right) = \dot{n}, \tag{7.15}$$

where D is the effective diffusion coefficient. Similarly, we have for the energy balance

$$\lambda_s \nabla^2 T = \dot{n}\Delta H, \tag{7.16}$$

where T is the local temperature, λ_s is the thermal conductivity, and ΔH the enthalpy change in the reaction.

These two equations are not easy to solve to obtain T and c as functions of position, since the reaction rate, $\dot{n}$, is a function of both temperature and composition. However, a useful and very general result can be obtained if we relinquish the aim of obtaining both T and c as functions of position and wish to know the relationship between T and c.

From equations (7.15) and (7.16), it is clear that

$$D\nabla^2 c = \frac{\lambda_s}{\Delta H}\nabla^2 T = \dot{n}. \tag{7.17}$$

If we put

$$\alpha = D\Delta H/\lambda_s, \tag{7.18}$$

and we assume that α is, with satisfactory accuracy, independent of T, then from equations (7.18) and (7.17) we have

$$\nabla^2(\alpha c - T) = 0. \tag{7.19}$$

If the concentration and temperature are both constant at the pellet outer surface, where they possess values c_s and T_s, then the solution of equation (7.19) is

$$(T_s - T) = \alpha(c_s - c), \tag{7.20}$$

as was first derived, for the particular case of a spherical pellet, by Damköhler [**18**].

We can now use equation (7.20) to estimate the *maximum possible* temperature difference between the interior of the pellet and its surface. This will clearly be obtained by putting c equal to its lowest possible value, i.e. zero for any reaction which is virtually irreversible, or the value corresponding to thermodynamic equilibrium for any reversible reaction. Such calculations, by Prater [**17**], and by others, e.g. [**19**], have shown temperature differences of up to 60 °C. The biggest differences occur when α is numerically large, as equation (7.20) shows, and equation (7.18) indicates how α depends on D, ΔH and λ_s. The particles considered were of roughly 10 mm diameter.

Example 7.9

Coke is being burned from catalyst spheres, using air at atmospheric pressure for the combustion. The catalyst spheres are 5 mm in diameter and made of a material of thermal conductivity 0.35 J m^{-1} s^{-1} K^{-1}. The combustion liberates 3.4×10^8 J per kmol of oxygen consumed. The diffusion coefficient of oxygen in the catalyst is 5×10^{-7} m^2 s^{-1} at the temperature of combustion, 760 °C.

What is the maximum steady-state difference in temperature between the centre of a particle and its surface?

[*Answer.* 1.2 °C. This question is based on figures given in [**17**].]

The above treatment has shown how to estimate the maximum *steady-state* temperature difference between the surface and the interior of a catalyst particle. It may be mentioned that under *transient* conditions much greater temperature differences may be encountered [**20**], but this will not be further discussed.

The temperature distribution and reaction rate within a catalyst pellet are thus related, and we have now to ask what effect this relation will have on the catalyst effectiveness factor. Let us rewrite equation (7.20) as

$$T/T_s = 1 + \beta(1 - c/c_s), \tag{7.21}$$

in which

$$\beta = \frac{D(-\Delta H)c_s}{\lambda_s T_s}. \tag{7.22}$$

β is the dimensionless Prater temperature rise and will be positive for an exothermic reaction.

The magnitude (and sign) of β is an important consideration – clearly a zero enthalpy of reaction will lead to an isothermal catalyst. If β is non-zero, as will generally be the case, we have to estimate the effect of this on the reaction rate, and for this we need to know

$$\gamma \equiv E_A/RT_s, \tag{7.23}$$

which is the dimensionless Arrhenius activation energy.

If γ is small, the effect of the temperature change on the rate of reaction is small, while if β is small, so is the temperature change itself. It will be seen later that the product $\beta\gamma$ can be used, with fair accuracy, to describe the results, as Aris has shown [**15** p. 144].

These factors were considered by Weisz and Hicks [**21**], whose results can be illustrated by the example shown in Fig. 36, which shows the effectiveness factor η. The following can be noted:

(*a*) If β is negative, the reaction is endothermic, and the effectiveness factor is *lower* than for the isothermal case;

(*b*) if β is zero, the reaction is isothermal. For this value only of β the effectiveness factor is independent of the value of γ;

(*c*) if β is positive, the reaction is exothermic, and the effectiveness factor is *higher* than for the isothermal case;

(d) if β is positive and large, the effectiveness factor can apparently, for a range of ϕ, have three possible values. This is an 'ignition' situation which the reader will better understand after reading Chapter 9.

It is convenient to have available a criterion capable of showing whether or not diffusion within a catalyst is likely to be a limiting factor. If the effectiveness factor, η, is greater than 0.95, then for practical purposes intra-particle diffusion effects can be neglected. Fig. 35 shows η as a function of ϕ, but it will be noted that ϕ is defined in terms of quantities, such as k, which cannot be predicted. In fact, if we are using a laboratory reactor to try and determine these quantities, the arguments can become circular unless care is used.

We can measure the reaction rate, $\dot{N}$ kmol s^{-1}, in a given reactor of volume V. This rate is given by

$$\dot{N} = k\sigma V(1-\epsilon)c_s^m \eta, \tag{7.24}$$

in which k is based on catalytic surface area, and m is the order of reaction. The Thiele modulus, ϕ, is given by equation (7.6), which can be written

$$\phi^2 = kc_s^{m-1}\sigma d^2/D. \tag{7.25}$$

Fig. 36. Effectiveness factor η versus Thiele modulus ϕ for various values of β (adapted from Weisz and Hicks [21]).

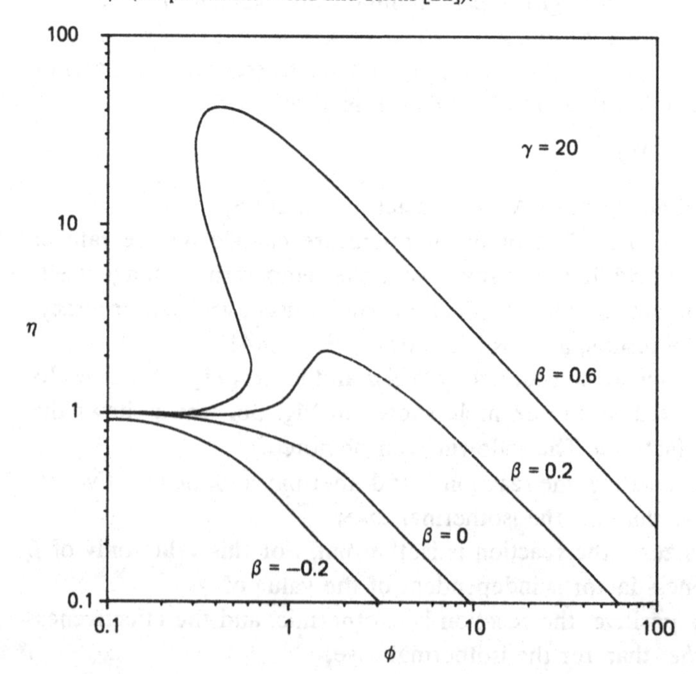

Eliminating the unknown k between equations (7.24) and (7.25) we have

$$\Phi = \eta\phi^2 = \frac{\dot{N}}{V(1-\epsilon)} \cdot \frac{d^2}{Dc_s}. \tag{7.26}$$

The parameter Φ contains quantities which can be observed (except D, which will usually have to be estimated as in § 7.3). The quantity d is the characteristic length of the particle, which is $R/3$ for a spherical pellet. Weisz and Hicks give as criteria for the absence of significant thermal or mass-transfer effects

$$\phi < \tfrac{1}{3} \exp[-\beta\gamma/2(1+\beta)].$$

or

$$\Phi = \eta\phi^2 < \tfrac{1}{10} \exp[-\beta\gamma/(1+\beta)]. \tag{7.27}$$

(These criteria are modified from those in [**21**] to account for the changed choice of d in equation (7.26).)

Since η is a function of ϕ, it is possible to present η as a function of $\eta\phi^2$, and Fig. 37 shows such a plot calculated for an isothermal spherical particle with first-order reaction. For this we have

$$\eta = \frac{1}{\phi}\left(\coth 3\phi - \frac{1}{3\phi}\right), \tag{7.28}$$

which is equation (7.13) when ϕ is defined on the basis of $d = R/3$. Fig. 37 will be acceptable for other geometries and orders, by analogy with Fig. 35, to which it is somewhat similar.

If Φ is less than 0.1, we can neglect mass-transfer effects, and this is what equation (7.27) will give for $\beta = 0$. We may now consider what β and γ are likely to be, and the consequences for the criteria of equation (7.27).

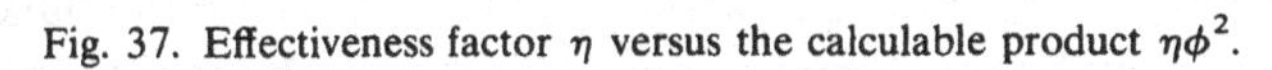

Fig. 37. Effectiveness factor η versus the calculable product $\eta\phi^2$.

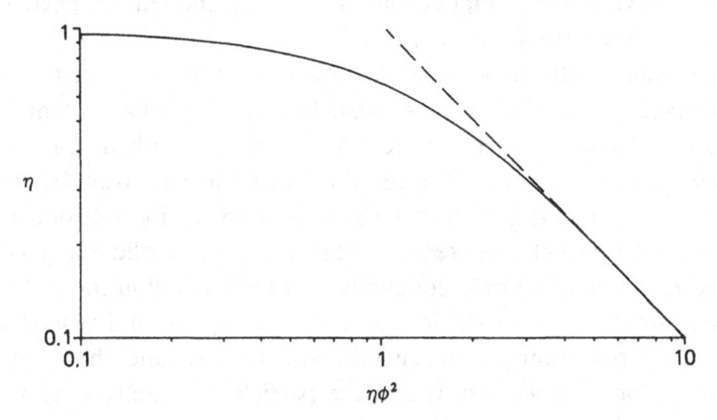

For gas–solid reactions, Dc_s is roughly constant with pressure, except at low pressures, when this product will fall. D will be somewhat less than the bulk fluid value (see equation (7.4)) and λ_s can be estimated for a porous solid. Putting an upper limit of 3×10^8 J kmol^{-1} for $(-\Delta H)$, Weisz and Hicks estimate that β will not be larger than unity, being usually somewhat smaller. For liquid–solid reactions, D will be some orders of magnitude smaller, and λ_s somewhat larger, so that β will be too small to give significant thermal effects. We note that if β is $\ll 1$, then the criteria in equation (7.27) depend only on the product $\beta\gamma$, as suggested earlier [**15** p. 244].

The value of γ might well be as high as 100, leading to values of $\beta\gamma$ up to about 10 for gas–solid reactions, but much less for liquid–solid reactions.

These conclusions must be treated with caution for more complicated kinetics. For example, Hutchings and Carberry [**22**] treat Langmuir–Hinshelwood kinetics with large surface coverages by both reagent and product. Again, diffusion effects can affect the *selectivity* of a reactor in situations where parallel reactions occur – see [**1**].

Finally, in this section, we must consider the possibility of a significant difference between the temperature of the surface of the catalyst pellet, T_s, and that of the bulk fluid flowing past it. This is the heat-transfer analogue of the mass-transfer problem considered in § 7.2, and is discussed in Carberry's textbook [**23**]. If heat transfer from the pellet to the bulk fluid is poor, the pellet may, for an exothermic reaction, be at a significantly higher temperature than the bulk fluid. This can be so even though the pellet may be at an acceptably constant temperature throughout its own volume. The existence of a glowing particle of coal in a fluidized combustor, surrounded by much cooler material, indicates that the effectiveness factor of a catalyst particle (based on the bulk fluid temperature) may be much higher than unity, even though the particle might internally be virtually isothermal.

In § 7.2 we came to the conclusion that mass-transfer resistance to the external surface of a porous catalyst particle is usually unimportant in comparison with that within the particle. Why is it that for heat transfer the position may be reversed? The reason is that for mass transfer the solid material of the catalyst hinders the mass transfer by diffusion of reagent within the particle, whereas for heat transfer the solid may have a much higher effective thermal conductivity than the fluid in the pores. This is particularly so for gas–solid reactions. For liquid–solid reactions the external heat-transfer resistance may also be low, and then temperature-difference effects both within the particle, and outside it, will probably be small.

***Example* 7.10**

A tubular reactor of volume 100 m^3 is packed with catalyst particles of equivalent radius 2.5 mm. The voidage is 0.4. When the reactor is in the steady state, 0.24 kmol s^{-1} of a gaseous reagent are decomposing according to an isothermal first-order irreversible chemical reaction. The effective diffusion coefficient of the reagent within the catalyst is 1.2×10^{-6} $m^2 s^{-1}$. The reagent is at 1 atm partial pressure and 700 K in the gas stream. Use equations (7.24)–(7.28) to estimate the effectiveness factor of the catalyst particles. (Fig. 37 can be used for a rough estimate.)

[*Answer.* $\eta \simeq 0.92$.]

***Example* 7.11**

A new catalyst material, for which $k\sigma$ is twice as large as in Example 7.10, will be used in the form of catalyst particles 2.5 mm in radius to carry out the same reaction to the same extent as in Example 7.10. Due to the different pore characteristics the effective diffusion coefficient is reduced to 7×10^{-7} $m^2 s^{-1}$. What volume of packing will be required? What size would the particles have to be to have the same effectiveness factor as in Example 7.10?

[*Answers.* About 60 m^3, 1.35 mm.]

***Example* 7.12**

Experiments with two sizes of catalyst particles are carried out in a laboratory investigation. The linear dimension of particles B is half that of particles A. What would you deduce from the following results for the reaction obtained per unit bed volume? Estimate effectiveness factors and Thiele moduli, if you can, using Fig. 35.

(i) Reaction B = Reaction A
(ii) Reaction B = 2 Reaction A
(iii) Reaction B = 1.5 Reaction A

[*Answers.* (iii) $\phi_A = 1.8$, $\phi_B = 0.9$, $\eta_A = 0.52$, $\eta_B = 0.78$.]

7.7 Catalyst formulation

When a solid catalyses a desired reaction it is often found that, even so, the rate per unit area of catalytically-active surface is small. Thus it becomes necessary to produce a high surface area of catalyst per gram if the catalyst is to be harnessed for the practical production of chemicals. Very small particle sizes are thus indicated – as the following example will show:

***Example* 7.13**

Spherical particles of carbon, of specific gravity 1670 kg m^{-3}, are to be used as a catalyst. Estimate the surface area per gram of carbon if the particles have radius 10^{-3}, 10^{-6} and 10^{-9} m.

[*Answers.* 1.8×10^{-3}, 1.8, 1800 $m^2\ g^{-1}$.]

A finely-divided 'activated carbon' may well have a surface area of 1000 $m^2 g^{-1}$, at which value it possesses sufficient reactivity to be of practical use. Considerations of containment of the particles, and of the costs of pumping fluid through the bed, rule out the use of such small particles unless they are compounded into pellets. These are large enough (say 5 mm in dimension) for it to be possible to pump the fluid through the bed at an acceptable pressure drop (or cost). The pellets must be porous, to provide access to the large surface area of the small particles making up the pellets, though the mass-transfer effects discussed in the previous sections may result in lowered effectiveness factors.

It is possible to model mass transfer in porous pellets by considering diffusion down cylindrical pores. We have not used this model, even though it forms an attractive, and tractable, mathematical proposition, because (*a*) the model is not physically realistic and (*b*) it provides results of the same *form* as those from the uniform continuum model (with D_{eff}) considered in § 7.4.

It is, however, instructive to estimate the (hypothetical) length of such pores in such a material.

***Example* 7.14**

An active porous carbon has a pore volume of $4\times10^{-7}\ m^3\ g^{-1}$. If it is assumed that this pore volume is made up of pores all with the same radius r, calculate this radius, and the total length of pores per gram, given that the active surface is 1000 $m^2\ g^{-1}$.

[*Answers.* $r=8\times10^{-10}$ m, total pore length 2×10^{8} km g^{-1}.]

The catalytically-active material may be so expensive (e.g. platinum) that porous pellets of this material alone are economically unacceptable. In such a case the catalyst may be finely dispersed on the surface of the pores of pellets of a cheaper 'support' material, such as silica, or alumina, or a blend. The support material may itself have some catalytic activity, but in other cases the support is inert.

Catalyst pellets must satisfy two physical requirements. They must be strong enough for the pellets at the bottom of a bed not to be crushed by the weight of those above, and they must maintain this strength under reaction conditions. With a tungstic oxide catalyst used for the hydration of propylene, polytetrafluorethylene was found to act as an effective 'binder' in the pelleting process, since it prevented the pellets crumbling in service to a fine dust, without reducing the reactivity of the catalyst.

'Promoters' may be added which, while not catalytically active themselves, improve the activity of the catalyst. They may act principally in maintaining the activity of a catalyst, for example by reducing the rate of sintering of small crystallites of catalyst (e.g. a metal on a support). The process of sintering of small crystallites together to form larger crystals reduces the active surface area.

The formulation of a suitable mix of particles of support, binder and promoter, the pelleting of such a mix, and the dispersion within the pellet of the (expensive) catalytically-active material can be seen to be a skilled business. The catalyst so made may be inactive when charged to the reactor. 'Activation' is then required. In some cases the reagent stream may itself activate the catalyst, in others a special treatment, e.g. with steam, or the reduction of metal oxides with hydrogen, may be required. In the latter case the catalyst may need special treatment on shut-down since, for example, many finely-divided metals are pyrophoric in air.

In practice one looks for the minimum charge of expensive catalyst and for the maximum possible duration before it has to be replaced. Catalytic activity is frequently a matter of comparative rates of desired and undesired reactions, and here a more selective catalyst may well be preferred to a more reactive, but less selective one. Yield and selectivity have been discussed before, in Chapter 6.

7.8 Laboratory testing of catalysts

It is essential for design purposes that there be experimental data on the performance of the catalyst in the form, and under the process conditions, in which it is actually going to be used. Many attempts to improve the catalyst formulation may be made and these have to be tested in the laboratory. Even on the laboratory scale such experimentation is time-consuming and costly. However, reduction in the amount of testing (to reduce these costs) should not be adopted if it results in data which are not sufficiently reliable for the purposes of full-scale design.

Firstly, the catalyst particles tested must be those actually to be used. The use of full-size catalyst pellets in small-bore laboratory tubular reactors can lead to deviations from plug-flow behaviour (the 'wall effect') but these defects are minor compared with those which can arise if experiments are performed with catalyst particles smaller than, or formulated differently from, those which will be used in practice, as is clear from §§ 7.4–7.7.

It is also important to use as reagent actual commercial material. This may well include impurities which could rapidly render useless a catalyst selected for its behaviour with analytical-grade material. We shall leave to the next section further consideration of catalyst deactivation.

Weekman [**24**] discusses the advantages and disadvantages of different types of laboratory reactors. We shall only consider one type of reactor, and shall assume that catalyst ageing is not rapid (though this type of reactor is better fitted than most for dealing with rapidly-deactivating catalysts).

A C.S.T.R. has the useful characteristic that conditions are uniform throughout the reactor, and its performance thus relates to a single set of reaction conditions. The P.F.R., on the other hand, gives a product resulting from a range of conditions from entry to exit. The problem is to produce C.S.T.R. behaviour in a laboratory reactor using industrial-scale catalyst particles.

A packed-bed reactor with a high recycle rate approximates to a C.S.T.R. if the conversion per pass is small. The flow rate is high, however, so the amount of reaction can be considerable, and can be determined by accurate analysis of the (small) fresh feed and product streams.

Two means of bringing this about are the spinning-basket reactor [e.g. **25**, **26**] and the pumped recycle reactor [e.g. **27**]. In the former the pellets are contained in a basket stirrer which agitates the fluid. In the latter, the catalyst pellets are contained in a fixed bed through which the fluid is pumped at a high (recycle) rate. Such reactors have also been used [e.g. **28**] with two-phase fluid reagent streams; in all cases it is necessary to ensure that the catalyst particles are uniformly 'available' to the reagent fluid.

If mass- and heat-transfer effects external to the particles are not important such a reactor will provide the kinetics of the catalyzed reaction for point values of reactor conditions, where the 'kinetics' will include the internal mass- and heat-transfer effects appropriate to those conditions. The design of a full-scale packed-bed reactor will then involve an integration, which will allow for the changes in conditions as the process stream passes down the bed. Though kinetic and transfer effects *within* the particles will have been properly modelled by the laboratory reactor, transfer effects *external* to the particles will not have been matched, since the laboratory reactor is probably operated at much higher Reynolds numbers than the full-scale reactor. Such external transfer effects, however, are often of little significance, but the point must be considered, especially with highly exothermic reactions.

7.9 Catalyst deactivation

Though there may be a short initial period during which the activity of a catalyst is increased in contact with the reagent stream, it will usually decline thereafter. This may require alteration of the reactor conditions to offset the declining reactivity. This subject will be discussed briefly under the following headings:

(*a*) the causes of deactivation;

(*b*) the consequences for operation of the reactor;

(*c*) the consequences for design and optimal operation of the process:

The reasons for the deactivation of catalysts are as diverse as the chemical reactions which it is desired to catalyze, and can be roughly divided into those which are physical and those which are chemical in origin. It may happen that under process conditions the catalyst 'ages' by a loss of surface area. This could be due to the sintering of crystallites into larger crystals; this may be prevented by the addition of promoters, as mentioned in § 7.8. Jensen and Ray [**29**] describe the effects of physical changes in supported metal catalysts; this follows some earlier work on unsupported metal catalysts. This type of surface change can lead to enhanced activity, and to oscillatory behaviour, but it is more common that surface physical changes will lead to a reduction of active area, and with that a reduction in activity.

The effect of side reactions may be to reduce the activity of the catalyst by 'fouling' its surface. This 'fouling' (if caused by the deposition of carbon it is called 'coking') may act physically, by obscuring the active surface, or by blocking pores for diffusion of reagent. In other cases the amount of material laid down may be very small, and the apparent area, and porosity, of the catalyst may be virtually unaffected.

Such a case may be described as 'poisoning'. In this case the catalyst surface is deactivated by a strongly chemisorbed layer of the poison, which may arise from an impurity in the feed, or be the product of some undesired reaction of the feed. If the poison is removed from the feed, activity may recover, as the chemisorbed poison is slowly removed by desorption or decomposition. Such reversible poisoning, perhaps arising from a temporary malfunction of the feed preparation, may be tolerable, if it is infrequent. If the poisoning is irreversible, it may be possible to restore activity chemically (e.g. by hydrogen reduction of an oxide layer, or steam treatment of alumina). In other cases the catalyst may have to be replaced. One part per million of an impurity in the feed may be all that is required to 'kill' a catalyst. Some years ago one of the authors was working on a plant using a supported platinum catalyst. This had operated entirely satisfactorily for about two years, but then 'died' in a few days. Procedures of increasing severity were used to try to revive the catalyst, but they proved unsuccessful. The catalyst had to be replaced, at a cost of over £100 000 – even after allowing credit for the recoverable platinum. The cause was never ascertained; the consolation was that the catalyst had lasted for over twice its design life.

The consequences for operation of the reactor depend firstly on the time scale of deactivation, and this will in fact have been an important consideration at the design stage. Let us for the present deal with the consequence of a deactivation taking place over a period of weeks, or longer. An example occurs in the isomerization of xylenes (to *p*-xylene, a starting material in the production of polyester fibres). Side reactions on the catalyst are of two types: disproportionation, to give lower-value products such as toluene and trimethylbenzene; and cracking reactions, which give rise to coking of the catalyst. The latter leads to a loss of activity, which may be compensated by operating at increased 'severity', by using higher temperature (and possibly higher pressure, too). Unfortunately such increased severity means lower selectivity – or else those conditions might have been originally chosen. The disproportionation reactions are favoured, by comparison with isomerization, when the severity is increased, and the coking reactions are even more favoured. Thus the rate of deactivation also increases. In due course the reactor has to be shut down and the catalyst regenerated, in this case by a controlled burning-off of the coke by nitrogen–air mixtures.

The consequences for design and optimal operation also depend in the first instance on the time scale. A catalyst deactivating over a year may be regenerated or replaced at the annual shut-down. A catalyst deactivating in days or weeks may be regenerated while its place in the process stream is taken by a duplicate reactor. If the catalyst deactivates in a shorter time, perhaps measured in seconds, then switching in of stand-by reactors becomes inconvenient or impossible. The first application of fluidized-bed reactors was to deal with the rapid deactivation of the catalyst used in the cracking of petroleum.

If the process of deactivation can be modelled, for example as a chemical reaction, then this opens up the possibility of mathematical optimization of the cycle of operation of a reactor. Szepe and Levenspiel [**30**] proposed this approach, and Levenspiel [**31**] devotes a chapter to this topic. Douglas and co-workers [**32**] discuss process design (and operating strategies) with catalyst deactivation.

These more quantitative approaches have not yet been widely adopted. This could be due to the variety of mechanisms of deactivation, and to the sensitivity of catalysts to poisons, which lays more stress on care to avoid poisons than on quantifying their effect.

It can finally be stressed that a catalyst which is selective, and 'robust', will always be chosen in practice over a catalyst which is more reactive initially, but which may be susceptible to poisoning or liable to favour side reactions.

7.10 The manufacture of ammonia

We shall conclude this chapter with a brief description of a typical plant for the manufacture of ammonia. We are grateful to the Agricultural Division of Imperial Chemical Industries PLC, for permission to reproduce a simplified flowsheet, Fig. 38, and to refer to their technical information, [**33**]. We particularly thank Dr G. Hargreaves and his colleagues for their co-operation and assistance.

The synthesis of ammonia is carried out on a very large scale. Plants, producing up to 1500 tonnes per day or more, are to be found throughout the world. The processes vary somewhat, particularly in the feedstock used to provide the necessary hydrogen, and in the operating pressure of the synthesis loop, but all make use of catalysis in several ways. The following example is that of a medium-pressure process using a natural gas feedstock, and is laid out in nine sections, in each of which a brief guide is given to the catalysis involved. Fig. 38 gives a simplified flowsheet, and the process will be considered in sections, as follows:

(1) *Hydrodesulphurization*

The feedstock consists of nearly pure methane, but contains small quantities of mercaptans. These are converted to hydrogen sulphide by reaction with hydrogen at about 400 °C and 20 bar over a catalyst consisting of mixed oxides of cobalt and molybdenum on alumina in the form of extruded cylinders some 3 mm in diameter. The *amount* of reaction is small, and the reactor is effectively isothermal. The catalyst is partially converted to sulphides, but they are at least as catalytically effective as the original oxides. If any oxides of carbon are present, the catalyst will act as a methanation catalyst (see (8) below) and may overheat. In normal operation this should not happen, and the catalyst then has a virtually unlimited life.

(2) *Sulphur removal*

The H_2S in the process stream is then removed by passing through a bed of zinc oxide pellets. This is not a catalytic reaction, but a reaction

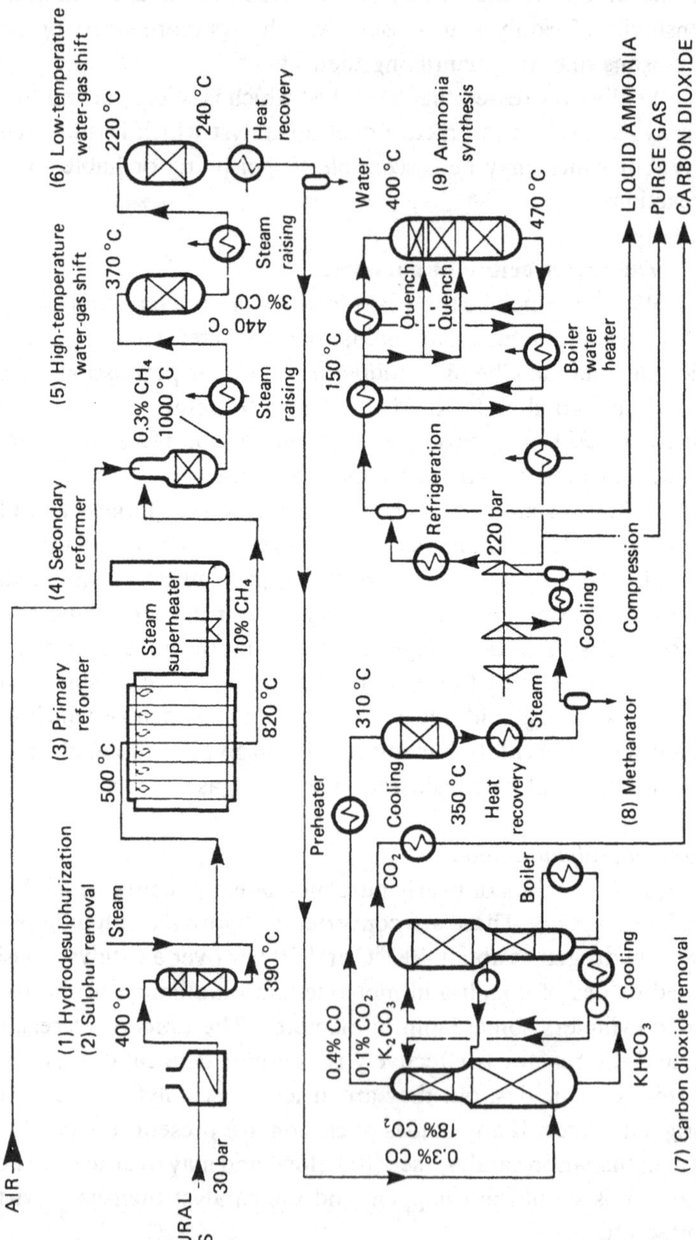

Fig. 38. Simplified flowsheet for typical ammonia plant. (By courtesy of Imperial Chemical Industries PLC.)

between the process stream and the solid packing. When the oxide is converted to sulphide, it is replaced. It is essential that the sulphur should be virtually completely removed, so that catalysts downstream are not poisoned.

(3) *Primary reforming*
Steam is added to the methane stream and the mixture is heated to about 800 °C in tubes fired from outside in a furnace. The reaction produces hydrogen and carbon oxides, and is highly endothermic. The tubes contain pellets of a nickel oxide catalyst on ceramic support; the oxide is reduced to nickel initially. These pellets are rings and are somewhat larger than most of the other catalysts (over 10 mm in size); they are contained in narrow-bore tubes (some 100 mm in diameter). This arrangement, which would not be considered ideal for a plug-flow reactor, is to assist heat transfer from the tube wall to and across the tube contents. The furnace operates at temperatures close to the metallurgical limits of the tubes. The prime aim is to get the heat into the process stream to encourage the reaction to go as far to completion at as high a temperature as possible. The exit stream contains some 7–10% methane unconverted.

The catalyst is temporarily poisoned by sulphur, but its activity is restored by further clean feed. It is permanently poisoned by other materials such as arsenic or chlorine, which can appear in the process stream from malfunction of certain recycle streams.

(4) *Secondary reforming*
Air is added to the product from the primary reformer in order to provide the nitrogen necessary for ammonia synthesis. The mixture is then passed over another catalyst bed, containing a somewhat similar catalyst to that in the primary reformer. The reaction of hydrogen with the oxygen of the air means that this reactor is exothermic overall, and the ceramic support of the nickel catalyst may have to withstand temperatures up to 1000 °C. As well as the removal of the oxygen, the methane content is further reduced, the aim being to lower it to about 0.3% or less.

(5) *High-temperature water-gas shift*
The process stream now contains hydrogen, nitrogen, water, carbon monoxide, and some carbon dioxide, together with argon from the air, and residual methane. Further steam is added, and the mixture passed over a catalyst of porous pellets of iron oxide promoted by chromium oxide. Process conditions are about 10 bar and 400 °C. The reaction is slightly exothermic and is diffusion-limited (the effectiveness factor is

low). The catalyst is poisoned by boiler solids in the steam, and its activity may also be reduced due to the effect of sintering at higher temperatures. The exit stream contains about 3% carbon monoxide.

(6) *Low-temperature water-gas shift*
The process stream is cooled to about 220 °C and is passed over a catalyst of mixed oxides of copper, zinc and aluminium; the copper oxide is reduced to copper in an initial activation. This catalyst is readily poisoned by compounds of sulphur and chlorine, and is very prone to sintering if overheated (for instance if the high-temperature shift reactor passes on too much carbon monoxide). As a result of the reaction $CO+H_2O \rightarrow CO_2+H_2$, the carbon monoxide in the exit stream should have been reduced to 0.3% or less. After cooling to atmospheric temperature, and separation of the condensed water, the product is passed to the carbon dioxide removal plant.

(7) *Carbon dioxide removal*
The removal of carbon dioxide from the cooled process stream has been carried out in a number of different ways during the history of ammonia production. One current method is to absorb the CO_2 in potassium carbonate solution, the reaction being homogeneously catalysed by arsenite ion. The CO_2 is subsequently removed by steam stripping and the potassium carbonate returned to the absorber. The CO_2, when purified, is a valuable by-product, being used, for example, to carbonate beer and soft drinks.

(8) *Methanation*
The process stream now contains about 0.1% CO_2 and 0.4% CO, which must be removed down to about one part per million. This is done by passing over a catalyst of nickel on alumina support at about 300 °C. The reaction of the carbon oxides with hydrogen is the reverse of the reforming reaction first carried out, and methane is formed. This methanation reaction is very rapid and also very exothermic; its purpose is to protect the ammonia synthesis catalyst from carbon oxides which would destroy its activity. The methanation catalyst is itself easily poisoned by sulphur, chlorine and arsenic compounds.

Malfunction of the earlier processes resulting in excessive concentrations of carbon oxides can cause dangerously high temperatures in the methanation reactor. The temperature can rise very rapidly and if this is sensed the process stream is bypassed to the atmosphere to prevent damage to the methanation reactor and its contents. Another reason for desiring low carbon oxides in the feed to the methanator is to reduce losses in the purge stream from the ammonia-synthesis loop.

(9) *Ammonia synthesis*

At long last we have a process stream containing hydrogen and nitrogen in the right ratio for ammonia synthesis. This 'synthesis gas' is compressed to about 220 bar ('high-pressure' processes work up to 1000 bar, 'low-pressure' processes at about 150 bar). The carbon oxides have been reduced to the order of one part per million. Argon is still present, from the air which provided the nitrogen, and there is some methane. To remove these from the system there is a small purge stream from the synthesis loop.

The synthesis of ammonia by passing hydrogen and nitrogen over an iron catalyst, the Haber process, has been operated for well over fifty years, and the chemical kinetics are still a matter for study and debate. The catalyst is formed by fusing magnetite with promoters such as oxides of potassium, aluminium and calcium. The cooled solid is broken up, graded and charged to the reactor. The circulation of hot synthesis gas reduces the catalyst to its active form of porous metallic iron, in which state it is pyrophoric in air. Providing sulphur has been removed from the feed, and provided carbon oxides are kept to about one part per million, the catalyst has an active life measured in years. The reaction is reversible and exothermic and hence the equilibrium yield falls with increasing temperature. The reactor is therefore designed with intricate internal heat exchange to approach the optimum rate of reaction. This is discussed in Chapter 9.

The efficient manufacture of ammonia, primarily for the production of fertilizers, is thus seen to involve some eight different catalytic steps. The plant manager must be aware of the sensitivity of these catalysts to poisons, or to undesired reagents. The formulation of the catalyst particles must take into account the mass-transfer and heat-transfer characteristics of the different reaction steps. Though practice is by now well established, the synthesis of ammonia still offers opportunities for improvement which, though apparently small in terms of efficiency, amount to large sums of money when the output from an individual plant may be 300 000 tonnes per year. These improvements are being brought about by the combination of the efforts of catalytic chemists and chemical engineers, and the purpose of this section has been to show how chemical reaction engineering is involved in these efforts.

Example 7.15

A zero-order isothermal reaction takes place on the internal surface of a catalyst pellet. If the pellet contains pores of radius r and length L, open at one end, determine the concentration of reagent as a function of distance from the pore mouth, and the effectiveness factor of the pellet. Show that an appropriate Thiele modulus for this geometry is $\phi = L\sqrt{(2k/rDc_0)}$, and that if $\phi > \sqrt{2}$ then the concentration of reagent is zero in the pores for distances $z > \sqrt{2}L/\phi$. In the latter case, what is the effectiveness factor?

Example 7.16

Small catalyst pellets, of average activity a_2, are fed to a C.S.T.R., in which they spend a mean residence-time of t_1. The leaving stream of partially deactivated catalyst has a mean activity of a_1. The pellets are crushed and the mixed particles are re-pelletized. They are then fed to a P.F.R., in which they spend time t_2. The rates of activation and deactivation of pellets obey the batch equation

$da/dt = \pm k$ (+for activation, −for deactivation).

The particles from the plug-flow reactivator have an activity of a_2, and are fed back to the C.S.T.R. The optimal design for the catalyzed reaction in the C.S.T.R. requires $kt_1/a_2 = 0.5$. Calculate t_2/t_1 for this to be the case.

[*Answer.* 0.865.]

Example 7.17

The catalytic cracking of a hydrocarbon has been studied using silica–alumina catalyst. Two sizes of spherical particles were examined: 0.1 mm diameter particles for a fluidized bed, and 5 mm particles for a packed bed. Estimate the effectiveness factor for these two catalysts, given that $D_{eff} = 1 \times 10^{-7}\ m^2\ s^{-1}$ and $k\sigma$ for this first-order irreversible reaction is $3\ s^{-1}$. Outline how you would determine k and σ in such an investigation.

[*Answers.* 1 and 0.20.]

Example 7.18

A first-order reaction is carried out in a packed-bed laboratory high-recycle reactor. From the following information deduce the effectiveness factor and the velocity constant, based on unit catalytic surface area.

Reaction rate $= 2 \times 10^{-5}$ kmol s^{-1} per m^3 of bed volume.
Fluid reactant concentration $= 0.6 \times 10^{-5}$ kmol m^{-3}.
Radius of spherical catalyst particles = 1 mm.

Porosity of particles = 0.46.
Crystalline density of catalyst material = 1750 kg m^{-3}.
Packed-bed density = 600 kg m^{-3}.
Catalytic surface area = 3×10^5 m^2 kg^{-1}.
Effective diffusion coefficient = 8×10^{-8} m^2 s^{-1}.

[*Answers.* $\eta = 0.14$, $k = 1.35 \times 10^{-7}$ m s^{-1}.]

Example 7.19

A gaseous substance A is converted to a gas B by a first-order irreversible reaction on the surface of a porous catalyst. B is itself decomposed to a worthless gaseous by-product C by a first-order irreversible reaction on the surface; the velocity constant of formation of B is four times that for its decomposition.

Set up the differential equations for the diffusion and reaction of A and B in a pore of length L and radius r, open to the bulk gas at one end, where the concentrations of A and B are a_0 and b_0.

Show that the concentration distributions of A and B along the pore are given by

$$\frac{a}{a_0} = \frac{\cosh \phi_A(1 - x/L)}{\cosh \phi_A}$$

and

$$\frac{b}{b_0} = \frac{\cosh \phi_B(1 - x/L)}{\cosh \phi_B}\left(1 + \frac{4a_0}{3b_0}\right) - \frac{4a_0}{3b_0}\frac{\cosh \phi_A(1 - x/L)}{\cosh \phi_A},$$

in which x is measured from the pore mouth, and the Thiele moduli are $\phi_A = \sqrt{(2k_1L^2/Dr)}$ and $\phi_B = \sqrt{(k_1L^2/2Dr)}$. It may be assumed that all species have the same diffusion coefficients.

Example 7.20

Using the results of Example 7.19, derive an expression for the yield of $B(= b_0/a_F)$ obtainable from a C.S.T.R. with a feed containing A only, at concentration a_F, which employs a porous catalyst modelled by pores of length L and radius r. It will be helpful to consider the fluxes of A and B at the mouths of the pores. Calculate the maximum value of this yield:

(a) when the Thiele moduli are both very small (no mass-transfer limitation). Comment on this answer, comparing with equation (6.17);

(b) when the Thiele moduli are both very large (severe mass-transfer limitations). Discuss the difference from (a).

[*Answers.* (a) 0.444, (b) 0.229.]

Symbols

c	Concentration, kmol m^{-3}.
d	Equivalent diameter of particle, m.
D	Diffusion coefficient, m^2 s^{-1}.
ΔH	Enthalpy change of reaction, J kmol^{-1}.
k	Reaction velocity constant.
k_m	Mass-transfer coefficient, m s^{-1}.
m	Order of reaction.
M	Molecular weight.
$\dot{n}$	Rate of consumption of reagent per unit volume, kmol m^{-3} s^{-1}.
$\dot{N}$	Observed reaction rate, kmol s^{-1}.
r	Pore radius or radial position, m.
R	Radius of particle, m.
T	Absolute temperature, K.
v	Superficial velocity through bed, m s^{-1}.
x, y, z	Co-ordinate distances, m.
α	See equation (7.18).
β	Dimensionless Prater temperature rise, see equation (7.22).
γ	Dimensionless activation energy, see equation (7.23).
δ	Stagnant film thickness, m.
ϵ	Voidage.
λ_s	Thermal conductivity, J K^{-1} m^{-1} s^{-1}.
μ	Viscosity, N s m^{-2}.
ρ	Density, kg m^{-3}.
σ	Surface area of catalyst per unit volume, m^{-1}.
τ	Tortuosity factor, see equation (7.4).
η	Effectiveness factor.
ϕ	Thiele modulus, see § 7.4.

References

1. Wei, J., *Ind. Engng Chem.*, 1966, **58**, Sept. p. 38.
2. Thoenes, D. and Kramers, H., *Chem. Engng Sci.*, 1958, **8**, 271.
3. Carberry, J. J., *A.I.Ch.E. Journal*, 1960, **6**, 460.
4. Mixon, F. O. and Carberry, J. J., *Chem. Engng Sci.*, 1960, **13**, 30.
5. Snowdon, C. B. and Turner, J. C. R., *Proceedings of the International Symposium on Fluidization*, ed. A. A. H. Drinkenburg (Netherlands University Press, Amsterdam, 1967).
6. Scott, D. S. and Dullien, F. A. L., *A.I.Ch.E. Journal*, 1962, **8**, 113.
7. Miller, D. N. and Kirk, R. S., *A.I.Ch.E. Journal*, 1962, **8**, 183.
8. Satterfield, C. N. and Sherwood, T. K., *The Role of Diffusion in Catalysis* (Addison-Wesley, Reading, Mass., 1963).
9. Satterfield, C. N., *Mass Transfer in Heterogeneous Catalysis* (M.I.T. Press, Cambridge, Mass., 1969).

10. Satterfield, C. N., *Heterogeneous Catalysis in Practice* (McGraw-Hill, New York, 1980).
11. 'Chemical Reaction Engineering, Nice', *Chem. Engng Sci.*, 1980, **35**, 3.
12. Thiele, E. W., *Ind. Engng Chem.*, 1939, **31**, 916.
13. Zeldovitch, Y. B., *Zh. fiz. Khim, U.S.S.R.*, 1939, **13**, 163.
14. Aris, R., *Chem. Engng Sci.*, 1957, **6**, 262.
15. Aris, R., *Elementary Chemical Reactor Analysis*, (Prentice-Hall, New Jersey, 1969, p. 136).
16. Aris, R., *The Mathematical Theory of Diffusion and Reaction in Permeable Catalysts* (Clarendon Press, Oxford, 1975).
17. Prater, C. D., *Chem. Engng Sci.*, 1958, **8**, 284.
18. Damköhler, G., *Z. phys. Chem.*, 1943, **A193**, 16.
19. Schilson, R. E. and Amundson, N. R., *Chem. Engng Sci.*, 1961, **13**, 226 and 237.
20. Wei, J., *Chem. Engng Sci.*, 1966, **21**, 1171.
21. Weisz, P. B. and Hicks, J. S., *Chem. Engng Sci.*, 1962, **17**, 265.
22. Hutchings, J. and Carberry, J. J., *A.I.Ch.E. Journal*, 1966, **12**, 20.
23. Carberry, J. J., *Chemical and Catalytic Reaction Engineering*, (McGraw-Hill, New York, 1976).
24. Weekman, V. W., *A.I.Ch.E. Journal*, 1974, **20**, 833.
25. Carberry, J. J., *Ind. Engng Chem.*, 1964, **56**, 39.
26. Patell, S. and Turner, J. C. R., *J. Sep. Proc. Tech.*, 1980, **1** (2), 31.
27. Bennett, C. O., Cutlip, M. B. and Yang, C. C., *Chem. Engng Sci.*, 1972, **27**, 2255.
28. Komiyama, H. and Smith, J. M., *A.I.Ch.E. Journal*, 1975, **21**, 664.
29. Jensen, K. F. and Ray, W. H., *Chem. Engng Sci.*, 1980, **35**, 241.
30. Szepe, S. and Levenspiel, O., *Chem. Engng Sci.*, 1968, **23**, 881.
31. Levenspiel, O., *Chemical Reaction Engineering* (John Wiley, New York, 2nd Ed, 1972).
32. Douglas, J. M., Reiff, E. K. Jr and Kittrell, J. R., *Chem. Engng Sci.*, 1980, **35**, 322.
33. *ICI Catalysts, Technical Information*, ICI Agricultural Division, Billingham, England.

8

Multiphase reactors

8.1 Mixing, mass transfer and kinetics

So far we have mainly considered reactors where only one fluid phase is involved, and where a solid, if present, acts only as a catalyst, i.e. is not itself a reactant. There are many cases, involving more than one reactant, where these reactants are supplied to the reactor in different, largely immiscible phases. It is then necessary that these phases be more or less intimately mixed, so that mass transfer of the reactants between the phases can take place effectively. When the reactants encounter each other at the molecular level (usually in solution in one of the phases) reaction will then take place according to some chemical kinetic scheme. These three steps of mixing, mass transfer and reaction may present widely-differing difficulties in different cases. When the reaction is 'slow', mass transfer may have a negligible effect on the overall rate. Conversely when the reaction is 'very fast', the overall rate will depend critically upon the method of mixing of the phases, and upon the mass transfer between them.

We have already considered, in the previous chapter, the packed-bed catalytic reactor. The flow phenomena and mass-transfer effects could be modelled usefully by the plug flow of fluid between particles and the homogeneous diffusion of reagent within particles (with an 'effective diffusion coefficient').

We must now consider more complicated systems where, for a start, there are at least two moving phases. We shall encounter gas–liquid, liquid–liquid, gas–solid and liquid–solid cases, and there are sometimes wide differences between reactor types in each of these divisions.

In recent years, an enormous literature has grown up, as each type of system has led to its own set of workers, techniques, terminology and textbooks. For example, at a recent conference on chemical reaction engineering, Van Landeghem [1] gave a plenary lecture with some 200 references, mostly between 1977 and 1980. An introductory text such as this can only hope to make the reader aware of these areas, and give the reader an entry into them.

8.2 Gas–liquid reactors: packed-bed absorbers

The absorption of gases into liquids is a unit operation of long standing. An important example, mentioned in § 7.10, is the separation of CO_2 from ammonia-synthesis gas, which in the past was carried out by absorption of the CO_2 into water under pressure, e.g. 50 bar. Similarly the recovery by absorption of components present as dilute mixtures with air is widely carried out, because either these components are valuable and/or they cannot be tolerated, on environmental grounds, in the effluent from a plant. As well as 'physical absorption' of gases into solution in a liquid, 'absorption with chemical reaction' is also employed and we shall see how the presence of the chemical reaction can lead to more effective absorption.

Gas absorption is commonly carried out in a column containing an inert solid packing, the liquid flowing down over the packing, and the gas flowing counter-currently upwards. This system possesses two advantages: firstly, a large interfacial contact area between the gas and liquid is obtained with the presence of packing (this is an important objective of the manufacturers of such packing); and secondly, the two phases pass in roughly plug flow through the column, which gives the maximum advantage of counter-current operation.

Correlations for the mass-transfer coefficients for 'gas-side' and 'liquid-side' resistance to mass transfer can be found in any book on gas-absorption. The gas-side mass-transfer coefficient k_G can be expressed by correlations of the form

$$Sh_G = \frac{k_G d}{D_G} = \text{const. } Re_G^{0.8} Sc_G^{0.5}, \tag{8.1}$$

where the subscript G indicates reference to the gas phase. This equation resembles equation (7.3) and comparable equations for heat transfer.

Mass-transfer coefficients on the liquid side are less well understood, but correlations of the form

$$Sh_L = \frac{k_L d}{D_L} = \text{const. } Re_L^{0.67} Sc_L^{0.5} Gr_L^{-0.5}, \tag{8.2}$$

where the subscript L indicates reference to the liquid phase, have been proposed.

The use of equation (8.1) or (8.2) for the purpose of calculating mass-transfer rates requires a knowledge of the concentration of the dissolving species at the gas–liquid interface. In general this is not available, and so the assumption is made that there is equilibrium at the interface. This introduces the solubility coefficient of the component (often expressed as a 'Henry's Law coefficient'), and enables the interface compositions to be eliminated. The result gives the 'overall mass-transfer coefficient' to be used with an 'overall driving force', which can be expressed in terms of gaseous concentrations (K_G) or liquid concentrations (K_L). Gas partial pressures are often used, which requires care in the dimensions of k_G and K_G.

In the absorption of gases into liquids we are usually concerned with substances of small or moderate molecular weight (since large molecular weights lead to low vapour pressures). For such gases the values of D_G do not vary very greatly from substance to substance. The same can be said for D_L, though values of D_L are vastly different from values of D_G. Thus for given fluid mechanics k_G and k_L in equations (8.1) and (8.2) are much the same for those species usually being considered.

The *solubility* of gases does vary very widely, however, and it is this factor which predominates in determining whether a given process will be 'gas-film controlled' or 'liquid-film controlled'. This is seen from the equation

$$\frac{1}{K_G}=\frac{1}{k_G}+\frac{H}{k_L} \tag{8.3}$$

which is readily derived and no doubt is familiar to the reader. A large value of H, the Henry's Law coefficient, is associated with low solubility and liquid-film control (e.g. O_2 desorption). A low value of H means high solubility and probably gas-film control (e.g. absorption of NH_3 from air mixtures).

The *capacity* of a liquid for physical absorption is determined by the solubility of the absorbed material. Many gases have low solubilities, and the capacity of the liquid can be increased if it contains a substance which will react with, and hence destroy, the absorbed component. An example might be the absorption of CO_2 into NaOH solution. On an industrial scale it is convenient if the absorbing liquid can be cheaply regenerated. In some cases this can be done by heating and reversing the absorption reaction; an example is the absorption of H_2S or CO_2 into aqueous ethanolamine, or of CO_2 into aqueous potassium carbonate/bicarbonate. The weakly acidic gas can then be driven off by heating the solution, which can thus be recycled to the absorption system after cooling.

'Absorption with chemical reaction' is usually regarded as an extension of gas-absorption theory rather than as part of reactor design theory. If the reaction is *fast* its effect is to reduce the liquid-film resistance. On the Whitman film model, the absorbed gas is destroyed by the chemical reaction within the film – none gets through into the bulk liquid. The effect of chemical reaction is that the measured 'mass-transfer' coefficient is increased. The capacity of the liquid is also increased.

If the reaction is slow, it has no effect on the mass-transfer coefficient, but the capacity of the liquid is greater than would be the case if there were no reaction. The driving force for absorption is also increased over that occurring for physical absorption. This is because the absorbed gas does not build up to the same extent in the bulk solution, but is destroyed by reaction. If the reaction is very slow, this effect may not be as useful practically as desired. An example is the reaction of CO_2 with CO_3^{2-}/HCO_3^- solutions, which can be greatly speeded up by adding arsenite ion, which catalyses the reaction (homogeneously). A review of CO_2 absorption discusses this point [**2**].

Chemical reaction may also be said to increase the *effective* interfacial area between gas and liquid. There are regions within the bed where the liquid is nearly stagnant and can become almost saturated with the absorbing gas. Thus such regions contribute little to the overall absorption rate. The presence of a chemical reagent in the liquid leads to an increase in the capacity of the liquid to absorb the gas, and thus regions which are nearly stagnant (and so almost saturated in the case of physical absorption) may have a sufficiently rapid through-flow to be effective if chemical reaction is also involved.

The subject of gas–liquid reactions is dealt with by Danckwerts in his book [**3**], which is essential reading for those working in this field. We shall look at one example only, which illustrates some points of general value.

This is the irreversible first-order reaction of a gaseous solute in a liquid, to give an involatile product. This might be, for instance, ammonia into a large excess of sulphuric acid in water. The 'large excess' is necessary, since the reaction is actually of higher order, and only becomes (pseudo)-first order if the acid concentration is nowhere significantly different from its value in the bulk liquid. We shall also consider the flow pattern to be such as to give a 'stagnant-film' in the liquid at the gas–liquid interface, i.e. the Whitman film theory is being assumed.

It may well be objected that the film model is physically unrealistic, and requires an arbitrary parameter – the liquid film thickness, δ. More sophisticated treatments do exist, e.g. the Danckwerts surface renewal theory; but this also requires an arbitrary renewal parameter, and, as Danckwerts himself points out [**3** p. 109], the different models in practice often give very similar results.

Consider a solute diffusing into a stagnant liquid film, as in Fig. 39. We will suppose that the concentration of the solute in the liquid at the gas–liquid interface is known. This will be the case if equilibrium with the gas is, as usual, assumed, and if the gas-film mass-transfer resistance is either negligible, or can be allowed for.

A material balance over the element dx will give the following equation:

$$D\frac{\partial^2 c}{\partial x^2}=kc, \tag{8.4}$$

in which we assume a steady state, (pseudo)-first-order kinetics (with velocity constant k) and constant diffusion coefficient, D.

This equation is similar to that required for Example 7.19, though the boundary conditions are not precisely analogous. It is helpful to convert equation (8.4) into dimensionless form (and this is necessary if numerical solution is required). Using the concentration ratio $\psi = c/c_i$, and the fractional distance into the film, $\xi = x/\delta$, we have

$$\frac{\partial^2 \psi}{\partial \xi^2}=M\psi \tag{8.5}$$

where $M = k\delta^2/D$. The boundary conditions are

$$\begin{aligned} \psi &= 1 \quad \text{at } \xi = 0, \\ \psi &= \psi_1 \quad \text{at } \xi = 1, \end{aligned} \tag{8.6}$$

where $\psi_1 = c_b/c_i$.

Fig. 39. Simultaneous diffusion and reaction in stagnant-liquid film.

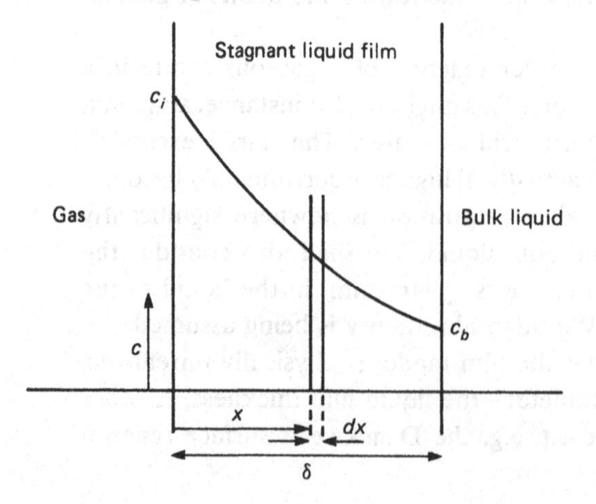

***Example* 8.1**

Show that the following equation satisfies equations (8.5) and (8.6)

$$\psi = \cosh\sqrt{M}\xi + \left(\frac{\psi_1 - \cosh\sqrt{M}}{\sinh\sqrt{M}}\right)\sinh\sqrt{M}\xi. \tag{8.7}$$

We note that the concentration gradient $\partial\psi/\partial\xi$ is negative throughout the film, and that the curve of ψ against ξ is concave upwards.

Though c_i and c_b are regarded as known, at a given position, P, in the column, they do actually depend on the extent of absorption elsewhere in the column, since the counter-current streams have been involved in absorption on their way to P.

***Example* 8.2**

If the liquid entering the column contains no solute, then diffusion of solute is always in the direction of ξ increasing. By considering the concentration gradient at $\xi = 1$, show that ψ_1 is less than $\cosh^{-1}\sqrt{M}$, and that therefore if $\sqrt{M}$ is large, ψ_1 is very small indeed.

As Example 8.2 shows, large M leads to negligible solute in the bulk liquid. In other words, all absorbed solute reacts in the stagnant film; none gets through to the bulk liquid. The parameter M is equal to $k\delta^2/D$, in which δ is the (apparent) film thickness, and is itself estimated from correlations for the mass-transfer coefficient k_L, see equation (8.2). The film theory says that $\delta \equiv D/k_L$, and thus it is more meaningful to express M as

$$M = \frac{kD}{k_L^2}. \tag{8.8}$$

It can, however, be seen from $M \equiv k\delta c_i/(Dc_i/\delta)$ that M is a measure of the ratio of chemical reaction in the film to diffusion through the film.

***Example* 8.3**

We have seen from Example 8.2 that if M is *large*, none of the solute gas gets through the stagnant film unreacted. Show that if M is *small*, then the concentration profile is a straight line, of slope $(\partial\psi/\partial\xi) = -(1-\psi_1)$, and that reaction in the film is negligible.

Though small M leads to negligible reaction in the film, the reaction in the bulk liquid may well mean that ψ_1 is not as large as it would have been in the absence of chemical reaction. Thus the effect of reaction is to increase the concentration gradient and so to increase the absorption rate. The *enhancement*, E, is the increase of the absorption rate in the presence of chemical reaction over that which would be obtained in the absence of chemical reaction, *but with the same values of* c_i *and* c_b.

Example 8.4

By considering $(\partial\psi/\partial\xi)$ at $\xi=0$, show that the enhancement, E, is given by $E=\sqrt{M}$, when M is large.

Fig. 40 shows diagrammatically the effect of increasing chemical reaction rate (e.g. increasing k) on the concentration profile at a given point in a packed column. Diagram (a) shows physical absorption, $M=0$, $E=1$; (b) shows the situation for small M. E, as defined, still equals unity, but c_i and c_b will both be lower (increased absorption elsewhere in the column lowers c_i, and reaction in the bulk liquid lowers c_b); (c) shows the situation for larger M. Now the concentration profile is curved, and E is greater than unity. Finally (d) shows the situation for $M \simeq 20$.

Fig. 40. The effect of increasing rate of chemical reaction on the concentration profile of reactant in the liquid film.

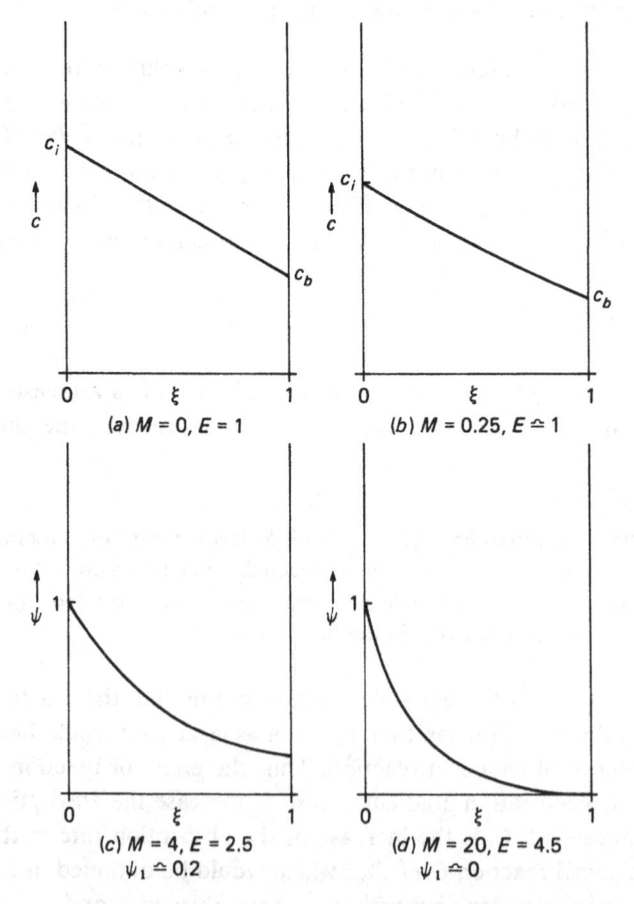

The enhancement is now approximately equal to $\sqrt{M}$ and ψ_1 is nearly zero. For increasing values of M, the profile steepens up, ψ being reduced to approximately zero at decreasing values of ξ.

It is more usual to absorb a gas into a liquid containing roughly stoichiometric quantities of a second reagent. Then the kinetics will not be pseudo-first order. An example is fast, irreversible, second-order reaction between two reagents. One diffuses into the film from the gas, the other diffuses from the bulk liquid into the film. Reaction takes place in a small region within the film. This problem was first examined by Hatta [**4**] in 1928, and is described in [**3**], which also discusses such factors as heat effects and ionization in absorbers. An interesting paper by Joosten *et al.* [**5**] considers the absorption of NO_X gases into nitric acid, where the gas and liquid films are both involved. There are many examples of different kinetics discussed in the literature; we have only given an introduction to the subject here.

8.3 Bubble-columns

We now consider gas–liquid reactors in which there is no solid packing, but where a vertical column contains a continuous liquid phase, into which gas is injected at the bottom. The gas rises through the liquid in the form of bubbles; mass transfer occurs from the gas to the liquid, where chemical reaction then ensues.

This arrangement is different from a packed column in two ways which can be advantageous. The liquid hold-up per unit reactor volume is higher, which gives a slow chemical reaction more time to occur (for a given liquid throughput). Secondly, the liquid may contain a dispersed solid, as with a biological culture, and a packed bed might rapidly be clogged by such a mixture.

The engaging simplicity of this apparatus is found to be more apparent than real as soon as its design is considered. There is an enormous literature on the subject – see, for example, [**1, 6, 7**]. Andrew [**7**] states that the literature, at least in so far as it applies to microbiological reactors such as fermenters, is 'not particularly informative as the authors invariably end up with empirical correlations which do little more than describe in symbols the obvious fact that more agitation usually produces more gas dispersion and more gas dispersion produces more mass transfer.' We shall here only attempt to describe some of the problems which arise.

The absence of packing in a bubble column introduces two severely complicating factors: (i) the presence of bubbles; and (ii) the existence of gross circulation patterns within the column.

(i) There are various ways in which gas can be introduced into the base of a bubble-column, for example through jets (tangential or otherwise), multi-orifice spargers or porous disc distributors. What size of bubbles is desirable? If too large, they rise rapidly through the liquid, and the time for mass transfer is small. If too small, they rise so slowly that, if the feed gas contains an inert component, most of the bubbles in the reactor will have lost their reagent and be effectively 'dead'. The coalescence of small bubbles into larger ones is generally undesirable, and is greatly accentuated by surface-active substances such as are always present in industrial reactors, even if avoidable in the laboratory. At the top of the liquid surface the bubbles should disengage easily. If they do not, but form a stable foam, this foam may be carried over to subsequent plant. 'Anti-foam' may thus be necessary, but too much causes coalescence in the bulk liquid, with loss of mass-transfer area and time. Andrew [**7**] gives advice to the cautious designer.

(ii) The introduction of gas at a low superficial velocity can lead to a uniform dispersion of small bubbles which rise through the liquid with little back-mixing. This 'pseudo-homogeneous' regime, [**1**], breaks down at superficial velocities greater than $0.1 \mathrm{~m} \mathrm{~s}^{-1}$ into a turbulent 'heterogeneous' regime, with much churning and backmixing of both phases.

Even a 'homogeneous' dispersion can change, as the result of a blockage of only a few (even one) orifices of the distributor, into a circulation pattern where the gas streams up in one region of the cross-section, dragging liquid with it. The liquid travels down elsewhere. For reliability of performance it may be preferable to introduce a 'draught tube' to encourage this liquid circulation. If the circulation is rapid, small bubbles may be carried round by the liquid, or be entrained from the layer of disengaging bubbles at the top.

If the distribution of gas, and the flow of both phases, can be reliably ascertained, the kinetics of the reaction in the liquid must then be considered. Sometimes these kinetics are of little concern – mass transfer is then all-important. If the reaction is simple practical problems may not arise, but frequently this is not the case. For example, in the use of air to oxidize hydrocarbons, or other organic chemicals, there are almost always competing reactions, which give undesirable, or worthless by-products. An example is the oxidation of cyclohexane to cyclohexanone and cyclohexanol. This mixture of ketone and alcohol is subsequently treated to give adipic acid, a precursor in the production of nylon. For high selectivity it is desired to stop the reaction product being further oxidized to worthless by-products. The reactor conditions are unfavourable to high selectivity, since the product is produced close to the bubble–liquid interface, where it is vulnerable to further oxidative attack before it can diffuse away into the bulk liquid. As a result, such reactors are operated at low conversion to obtain high selectivity. A consequence is the separation and recycling of large quantities of unreacted cyclohexane, which is very costly.

Advantage can be taken of liquid circulation, as indicated diagrammatically in Fig. 41. If air is introduced at *A* into a liquid then the difference of density will cause circulation as shown. When this has become established, it becomes possible to introduce air at *B*. If the bubbles are not large, they are carried down and round the cycle, disengaging at *C*. If the tube is tens of metres high, then the increased pressure at the bottom will aid solution of the feed gas. In fact the bubbles may dissolve entirely, re-appearing in the upward leg.

This has been used by Imperial Chemical Industries PLC in two biological processes, where the liquid is actually a suspension of bacteria, or of a yeast [**8**, **9**]. If air is injected into the downward leg, the gas

Fig. 41. Pressure-cycle reactor.

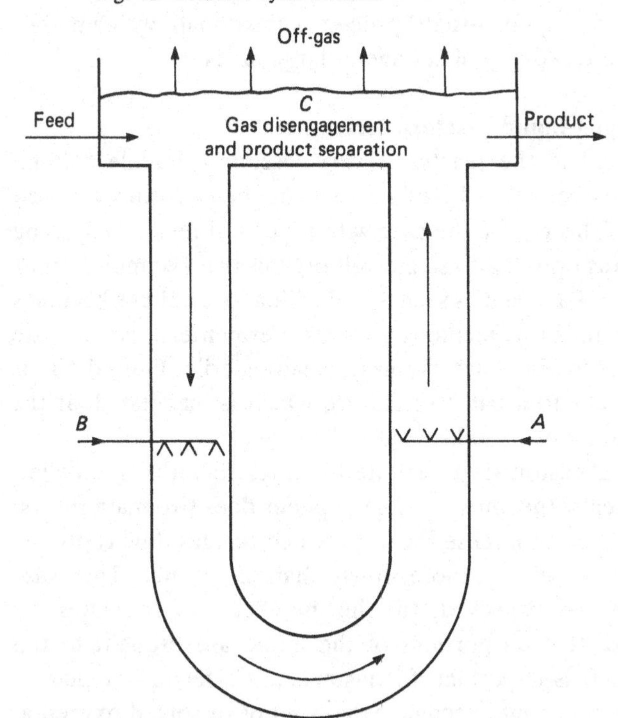

emerging in the upward leg removes carbon dioxide produced by the organism.

The two processes are (*a*) the production of 'Pruteen', a single-cell protein, and (*b*) the treatment of sewage. Their requirements are very different. For protein production the liquid-phase nutrient should be pure, and the aim is to maximize cell production (and thereby minimize CO_2 production). A carefully chosen organism is placed in the plant at the start, and it is 'harvested' from the top. It is essential to prevent the ingress of foreign organisms, which could, and almost certainly would, eliminate the chosen one. These requirements of biological hygiene are rather unfamiliar to the chemical industry, and require careful design of valves, pipelines, injectors, etc.

For the treatment of sewage, the liquid feed contains a wide, and probably changing, variety of organisms. In fact it is desirable that it should, since the non-biological components of sewage are themselves variable. Also the separation and disposal of the biological matter produced in the reactor is comparatively expensive, so here the aim is to *maximize* CO_2 production and minimize the production of 'sludge'. The upflow and downflow can be concentric, and can be contained in a concrete shaft buried in the ground below the separation unit. The ground area requirement of this 'deep-shaft' process is thus small, which makes it attractive for the treatment of sewage in large cities.

8.4 Agitated gas–liquid reactors, fermenters

If the liquid in the reactor is very viscous a bubble column becomes difficult to operate, and the liquid may not be uniformly exposed to the gas stream. This can be the case with industrial fermenters using fungal organisms (as opposed to single-cell organisms). Examples occur in the manufacture of antibiotics such as penicillin. The culture becomes a thick 'soup' containing filamentous growths. Fermenters also operate as semi-batch reactors, in which the gas (usually sterile, filtered air) is supplied continuously to a batch of culture which is 'harvested' at the optimal time for antibiotic yield.

In such cases it is customary to agitate the mixture with an impeller, releasing the air below the impeller. The impeller does two main things: (i) it provides a region of intense shear, in which the gas flow is broken into small bubbles, which rise more slowly than large ones. Thus gas–liquid mass transfer is improved; (ii) the impeller also circulates the liquid, and ensures that all portions of the liquid are brought to this high-shear region. It is important, for instance, in a fermenter that no regions of the culture should become exhausted of dissolved oxygen as the result of the metabolism of the organism.

Much academic work has been done using a single impeller in a tank of the same diameter as the liquid depth. In such tanks, the liquid is virtually completely mixed. On the industrial scale, such high inputs of energy per unit volume by the stirrer are expensive (and also lead to cooling requirements since the organisms are sensitive to temperature rise). Industrial fermenters often have different geometries, being much taller than their diameter, and using several impellers on a single shaft. Andrew [**7**] goes so far as to say that mass-transfer data from laboratory reactors are 'virtually useless for design purposes' (of full-scale plants).

On the full scale, it may be an acceptable simplification to say that virtually all the gas–liquid mass transfer occurs near the impellers, and that the circulation of the liquid (mean circulation time, and variance) then determines the reactor performance. Such circulation time distributions can be measured using a small, neutrally buoyant, radio transmitter as a 'tracer' [**10**]. The distributions can be related to stirrer type and speed, and can be used to determine optimal circulation rates.

8.5 Liquid–liquid reactors

The contacting of two immiscible liquids in a stirred tank is frequently used to extract a component dissolved in one phase into the other phase. Liquid–liquid extraction towers, or mixer/settler systems, are well-known in the oil industry. The major points of practical interest relate to the dispersion of droplets of one phase in the other, the maintenance of this dispersion for an adequate time for mass transfer to occur, and – most importantly – the separation of the dispersion when mass transfer has taken place.

Initial dispersion may be achieved through nozzles, or in a pump. One has to consider which phase should be dispersed in a continuous phase of the other. Too effective a dispersion may lead to an emulsion which is very difficult to 'break'; thus settler design may be more important than mixer design.

Mass transfer between the droplets and the continuous phase has excited considerable interest – see the references in [**1**]. Small droplets are 'immobile', by which is meant that they stay spherical even under the conditions of turbulence in the mixer, and mass transfer within them is by diffusion since there is little circulation in small droplets. The presence of surface-active agents may cause the droplets to remain 'rigid'. Larger droplets may have circulation within them induced by the flow outside (or even convection currents due to mass transfer inside). Still larger droplets may oscillate in the flow field, and be liable to break up into smaller ones. These latter phenomena greatly assist mass transfer between the phases. A spread of residence-times in the droplets passing through a mixer can be expected, and as there is also coalescence of smaller droplets, and break-up of larger ones, it is not easy to estimate the average concentration of the dispersed phase leaving the mixer.

Examples of liquid–liquid dispersions in which reaction also takes place occur in the case of polymerisation processes, and also processes such as the nitration of organic liquids. In the latter case, if the reaction is rapid, it may be hard to achieve a desired *mono*-substitution since further reaction of such a product close to the droplet surface may occur before it can diffuse into the interior of the droplet. This aspect of 'segregation' in reactors has been mentioned earlier, in Chapter 6.

Many polymerisation processes are carried out in suspension, for example the manufacture of polystyrene beads for ion-exchange resins. The droplets are typically of 0.5 mm diameter. The kinetics are those of bulk polymerisation – the suspension enables good temperature control (addition polymerisation is very exothermic) and also provides a product which is easy to handle and in a form to fit its eventual use. The kinetics of *emulsion* polymerisation, on the other hand, are different. Here the droplet size is 1 μm or less, and the initiator for the polymerisation is in the aqueous phase, not the organic phase. Additives, including soaps, are used, not only to stabilise the emulsion, but also to play a role in the polymerisation (see [**11**]). This process has been used to make styrene/butadiene copolymer, a synthetic rubber.

8.6 Fluid–solid reactions. Packed-bed absorbers

Packed beds are frequently used to remove a component from a fluid stream. This may be done by adsorption (as in packed-bed driers), or by exchange with a component in the bed (as in ion-exchange beds). Though the performance of such apparatus is usually more a matter of mass-transfer kinetics that of reaction kinetics, it is appropriate to consider them briefly here.

The removal of water vapour from a gas stream may be effected by passing the gas through a packed bed of silica gel or activated alumina. When the bed has absorbed water to such an extent that some water vapour escapes in the product ('breakthrough') the bed is taken out of the process for regeneration, and a second, regenerated, bed may be placed in line for the process to continue.

Regeneration may be done by passing hot air through the 'drier' to dry it. This procedure may also be effective in other cases, e.g. in the absorption of CO_2 from the atmosphere in submarines. In this case the concentrated CO_2 stream from the regenerator is dealt with separately.

Regeneration can also be performed by change of pressure. In 'pressure-swing absorption', one component, e.g. N_2, is absorbed preferentially at high pressure, and is released by passing gas, e.g. air, at low pressure through the bed. The product stream in that case is high-pressure air enriched in oxygen.

The commonest use of ion-exchange is in the demineralization of water, for example in the preparation of ultra-pure water for high-pressure boiler feed in power generation. In this case chemical regenerants are used to displace the absorbed, unwanted, ions, by hydrogen and hydroxyl ions. These when exchanging with the unwanted ions in the feed – for example sodium or chloride ions – produce water by subsequent reaction together in solution.

There are many general texts on absorbers, and on ion-exchange. It is necessary to know the equilibrium distribution of the solute between the phases, as well as the transfer rates. If the solid phase 'prefers' the solute (in ion-exchange, for example, sodium ions are preferred to hydrogen ions on the resin), then a sharp front may tend to develop in the bed. The rate of advance of this front can be calculated by material balances – it is a 'shock-wave' phenomenon. If the solid phase does not prefer the solute, a broadening front develops, which may break through when quite a high proportion of the theoretical capacity of the bed is still unused. The shape of this front within the bed at any time is not a simple matter to calculate. In both cases, preferred and unpreferred, the finite *rate* of exchange may lead to further broadening. Fig. 24 in Chapter 5 gives an illustration of these effects.

8.7 Fluid–solid reactions. 'Shrinking particles'

We are concerned here with cases where the solid phase is (irreversibly) converted by reaction with the fluid phase.

Examples occur widely in the mineral-processing industry, where the 'roasting' of ores is commonplace. The burning of coal, the operation of a blast furnace, the production of lime, the washing or dissolution of beds of crystals, and some electrochemical processes are further examples. In some of these cases an adequate description of the process inevitably involves simultaneous consideration of chemical reaction, mass transfer *and* heat transfer. We shall only consider the simplest situations, in which heat-transfer resistance does not affect the rate.

The reaction of a (roughly) spherical solid particle with a fluid stream may result in a reaction product which is soluble, or volatile, in the fluid stream. On the other hand the particle may become coated with a layer of porous ash, oxide, or similar material. In the latter case the reagent has to diffuse from the fluid stream to the surface of the unreacted solid, which we shall assume to be non-porous. This will be taken up in § 8.8.

In the first case the solid particle will shrink until it disappears. The process may be controlled by either

(*a*) mass transfer of reagent to the solid surface;

or

(*b*) chemical reaction at the surface.

(Mass transfer of product away from the surface could also be a controlling factor, though this is less likely. Examples occur in electrochemistry.)

Let us consider these two situations separately at first. The efficacy of mass transfer to the solid particle depends upon whether the particle is in a fixed or fluidized bed, and in general upon the fluid-mechanical situation. Equations such as

$$Sh = k_M d/D = ARe^{1/2}Sc^{1/3} \tag{8.9}$$

have already been discussed in § 7.2. As a shrinking particle dimishes in size, so Re will tend to zero, and modifications to equation (8.9) of the type

$$Sh = 2 + ARe^{1/2}Sc^{1/3} \tag{8.10}$$

have been suggested (e.g. **[12]** for fluidized beds). Here 2 is a theoretical result for mass transfer to a single sphere at $Re = 0$. Its applicability in equations such as equation (8.10) has been challenged **[13]**.

When chemical reaction at the surface is controlling, the treatment is simple. The reaction rate equation is

$$\dot{N} = kc^m \cdot 4\pi r^2, \tag{8.11}$$

where $\dot{N}$ is here the number of moles destroyed per second per particle, k is based on unit area, c is the concentration of reagent in the bulk-fluid phase, m is the order of reaction and $4\pi r^2$ is the particle surface area (we assume that the particle is non-porous). Since the particle is diminishing in size, $\dot{N}$ may also be expressed as

$$\dot{N} = -s \cdot 4\pi r^2 \, dr/dt, \tag{8.12}$$

where s is a constant, which involves stoichiometric coefficients and the density of the solid. Combining equations (8.11) and (8.12) we see that

$$dr/dt = -kc^m/s; \tag{8.13}$$

in other words the particles diminish in radius at a constant rate. (Note that equation (8.12) will hold whether chemical reaction *or* mass transfer is controlling, or if both of them are significant.)

Example 8.5

Zinc spheres 5 mm in diameter are to be dissolved in an acid solution. The process is controlled by reaction at the zinc surface, and experiments show that, at the particular acid concentration used, acid is consumed at the rate of 3×10^{-4} kg-equivalents of acid per second per m^2 of surface. How long will it take:

(*a*) for half the weight of zinc to be dissolved;

(*b*) for the zinc spheres to be completely dissolved?

The density of zinc is 7.1×10^3 kg m^{-3}, and its atomic weight is 65.4.

[*Answers.* (*a*) 373 s, (*b*) 1810 s.]

Example 8.6

Spherical particles of pure carbon are burning. The process is controlled by mass transfer of oxygen to the carbon surface. The mass-transfer coefficient obeys an equation of the form of equation (8.9).

What is the ratio of the time required for the radius of the particles to be halved to that required for complete combustion? Compare your answer to that if reaction at the surface were rate-controlling.

[*Answer.* 0.646. Compare 0.5.]

If *both* mass-transfer resistance and chemical reaction at the surface are significant for controlling the overall rate we can, if the reaction is first order, apply the treatment outlined in § 2.7. Equation (2.41) can be rewritten as

$$\dot{N}=\frac{kk_M}{k+k_M}\cdot c\cdot 4\pi r^2, \tag{8.14}$$

It is to be noted that the mass-transfer coefficient, k_M, is a function of the particle size and this, of course, is continually diminishing. This difficulty can be avoided by applying the 'quasi-static' approximation, so often used in similar situations. It is assumed that the particle size changes only slowly by comparison with the time required for steady states of diffusion to be set up. We then calculate the diffusional processes on the basis of constant particle size, surface area, etc., and use the results so calculated to determine the rate of change of the particle size. Of course, k_M must be allowed to vary with particle size, which makes it likely that the solution of equations (8.12) and (8.14) will involve a numerical or graphical integration to obtain r as a function of time.

The 'quasi-static' approximation requires checking in each case. If it is not a reasonable approximation, the problem becomes much more difficult to handle, and numerical methods will almost certainly be necessary.

8.8 Fluid–solid reactions. 'Ash-coated' particles

It may occur that, as a result of reaction, the solid particles become coated with reaction product. As the reaction proceeds the particle, in cross-section, appears as in Fig. 42. The unreacted core diminishes in radius, the layer of product (which may be a porous ash) becoming thicker, until finally the particle is completely converted to 'ash'. We shall assume that there is no volume change on reaction. When there is such a volume change this will probably result in the 'ash' layer either splitting (if its volume is less than that of the original particle) or cracking off in flakes (if its volume is greater, which frequently occurs with the rusting of iron). In either case 'fresh' surface is revealed and the following treatment is not applicable.

Assuming that Fig. 42 represents the process, there is the further possibility that reaction is controlled by diffusion of reagent through the 'ash' layer. This layer may be porous, or it may not; provided reagent can diffuse through it to the unreacted solid, the reaction will proceed. A good example of the contrary situation is provided by aluminium. This is a highly reactive solid, but nevertheless at temperatures which are not too high it rapidly coats itself with a thin, highly impervious layer of oxide and the oxidation reaction ceases.

Where an 'ash' layer is present it is very unlikely that mass transfer to the outer surface of the particle will affect the reaction rate, since mass transfer through the layer will be much more difficult. The reaction rate may therefore be affected by diffusion through the ash, by chemical reaction at the surface of the unreacted solid, or by a combination of the two.

If chemical reaction at the surface of the unreacted solid controls, the situation is the same as in the previous section. The 'ash' might as well not be there, so far as the overall reaction rate is concerned.

If diffusion through the 'ash' layer controls, the quasi-steady state treatment is straightforward. Let it be assumed that the flux of reagent towards the unreacted solid is given by

$$N_A = D\frac{\partial c}{\partial r}, \tag{8.15}$$

Fig. 42. The model for reaction of a spherical particle to 'ash'.

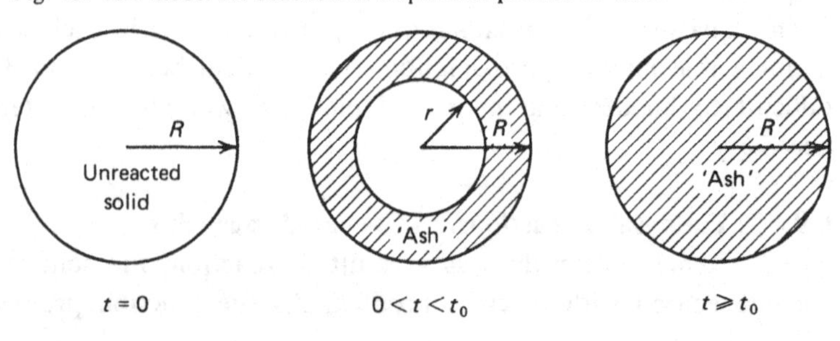

in which D is an effective diffusion coefficient. (This assumption would require justification in any given case.)

For diffusion through a layer of 'ash' on a spherical particle, if the concentration of reagent is c_s at the outer surface, and effectively zero at the reaction surface, it is easy to show that

$$\text{Reagent entering particle} \equiv \dot{N} = 4\pi D c_s \frac{Rr}{R-r}. \tag{8.16}$$

At the start of the process the reaction must be chemically controlled, since equation (8.16) would give an infinite mass transfer when $r \to R$. Later, as the 'ash' thickens, so the mass transfer resistance increases.

Using equations (8.16) and (8.12) we obtain

$$\frac{dr}{dt} = -\frac{Dc_s}{s}\frac{R}{r(R-r)}. \tag{8.17}$$

***Example* 8.7**

Derive equation (8.16).

***Example* 8.8**

For a particle which reacts under 'ash-diffusion' control, what is the ratio of the time required for the radius of unreacted solid to be reduced to $R/2$ to that required for complete reaction? Compare with the answers to Example 8.6.

[*Answer.* 0.5.]

The reader who has worked out the examples will have available the three different expressions for the radius of unreacted solid as a function of time. They are:

(a) Shrinking particle, chemical reaction control, radius diminishes linearly with time, equation (8.13).

(*b*) Shrinking particle, mass transfer control given by equation (8.9)

$$1-\left(\frac{r}{R}\right)^{3/2} = \left(\frac{t}{t_0}\right), \qquad \text{cf. [\textbf{12}].}$$

(*c*) Ash-coated particle, chemical reaction control, as for (*a*).

(*d*) Ash-coated particle, ash-diffusion control,

$$1-\frac{3r^2}{R^2}+\frac{2r^3}{R^3} = \left(\frac{t}{t_0}\right).$$

These three curves are shown in Fig. 43. The time scales have been 'normalized' by use of t_0, the time required for complete reaction.

For the reaction of a given solid the chemical reaction rate constant and the effective ash-diffusion constant are experimental quantities to be determined. Mass transfer to the solid particle can be estimated from equations such as equation (8.9). Given this information it can be determined whether any one of the three processes dominates, or whether a combination of two of them may be controlling [12].

8.9 The physics of fluidized beds

A book on reactor design can hardly ignore the existence of fluidized-bed reactors. On the other hand there are many aspects of fluidization which are not yet understood, and the application of fluidization to the production of chemicals raises a further host of problems. It is impossible here to cover the different treatments of this huge field. We shall briefly note some of the problems and refer the reader to various textbooks, each with its distinctive approach to the subject.

We must first consider how a fluidized bed can be produced. If a fluid is passed through a packed bed of solid, then the pressure drop/velocity relation is as shown in Fig. 44. As the velocity increases, the pressure drop rises linearly at first but eventually the pressure drop across the bed will become equal to the (apparent) weight of the bed and the latter then tends to lift. If the particles are fairly small this will occur while

Fig. 43. The radius of unreacted solid as a function of reduced time (based on Yagi and Kunii [12]).

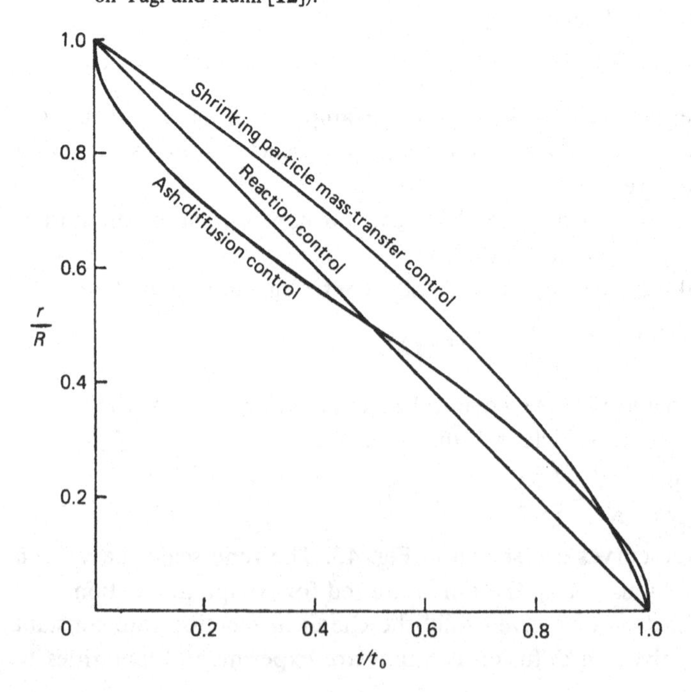

the pressure drop is still proportional to the velocity, and it occurs at the 'incipient fluidizing velocity'.

At higher velocities the pressure drop increases relatively slowly. In the region of the incipient fluidizing velocity, u_0, the behaviour shows hysteresis, in that the pressure drop depends upon the recent past treatment of the bed. This is also reflected in the height of the bed; a bed which has been fluidized, and through which the velocity is reduced, will settle gently. When it becomes still, and if the velocity is then lowered further, the bed may sink somewhat, especially if it is tapped or vibrated. This 'packing' is associated with an increase of pressure drop for a given flow.

At velocities higher than u_0, three types of behaviour can be described. Some materials 'do not fluidize well'; they tend to form aggregates. The distribution of fluid through the bed may be irregular and there may be stagnant regions. If the solid particles are of roughly spherical shape, and are not 'sticky', they will probably 'fluidize well'. In these cases, the behaviour at velocities higher than u_0 falls into two categories. If the fluid is a liquid, the bed expands by the particles moving further apart in a uniform manner. (This is generally called 'particulate' or 'smooth' fluidization.) At high enough velocities the individual particles cannot settle and they are entrained into the liquid stream.

Fig. 44. Pressure drop/velocity graph, showing incipient fluidization at u_0. Results in the region of 'hysteresis' depend on the recent past history of the bed.

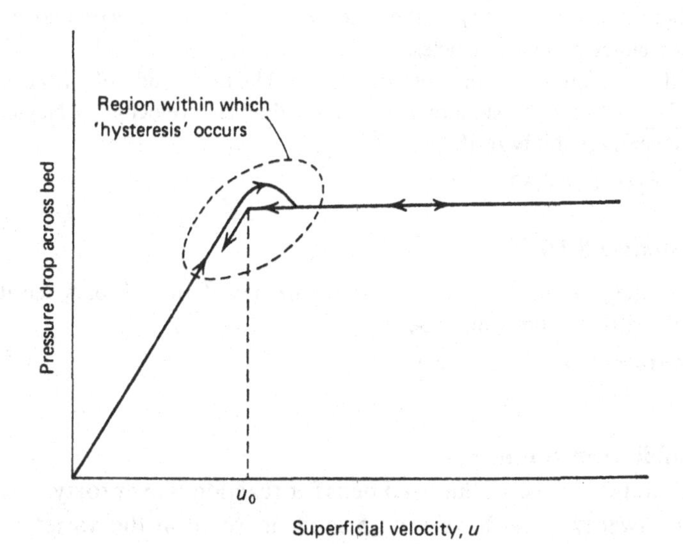

If the fluid is a gas, increase of u above u_0 produces 'bubbling' phenomena. The 'extra' gas passes through the bed as somewhat irregular bubbles which can move at speeds much greater than the mean velocity. These burst at the surface of the bed, which thus resembles a boiling liquid. This is called 'aggregative' or 'bubbling' fluidization. Those regions of the bed where bubbles are not present at a given instant have roughly the same voidage, and throughflow of gas, as at incipient fluidization. It is possible to obtain non-bubbling gas-fluidized beds, and bubbling liquid-fluidized beds, but these can be regarded for our purposes as curiosities.

The physical behaviour of actual fluidized beds may not obey at all closely the model behaviour of laboratory beds of uniform spherical particles. Factors such as the shape, size distribution, and stickiness of the particles may cause wide variations in behaviour. The presence of baffles or other internal fixtures within the bed may further modify the behaviour. If the bed is narrow, bubbles tend to 'bridge', or 'slugs' are formed. Furthermore, in industrial reactors, where the gas velocity may be a hundred times u_0, the bed resembles a violent sandstorm and the behaviour is not likely to be describable in terms of simple bubble fluid mechanics.

***Example* 8.9**

Resin spheres of diameter 0.9 mm and of density 1200 kg m^{-3} form a bed 1 m deep and of voidage 0.48. When fluid is passed through the bed the pressure gradient approximately obeys the equation

$$\Delta P/L = 460\mu u/d^2,$$

where μ is the viscosity of the fluid, u its superficial velocity and d is the diameter of the particles.

If the fluid is air, of density 1.3 kg m^{-3} and of viscosity 1.7×10^{-5} Ns m^{-2}, estimate the minimum fluidizing velocity, u_0. Neglect the compressibility of the air.

[*Answer.* $u_0 = 0.63$ m s^{-1}.]

***Example* 8.10**

If the bed in Example 8.9 is fluidized with water, of viscosity 10^{-3} Ns m^{-2}, what will u_0 be?

[*Answer.* $u_0 = 1.8 \times 10^{-3}$ m s^{-1}.]

8.10 Fluidized-bed reactors

Although the use of fluidized beds for reaction is over forty years old, the last twenty years have seen a great increase in the variety of reactions which are carried out in this manner. The solid particles which are fluidized may act as a catalyst for the reaction of material in the fluid stream, as in the 'cat-cracker' of the petroleum industry, or may themselves be converted by the reaction, as occurs in certain ore-roasting processes. The basic design problem in either case is the calculation of the mass transfer rates of reagent from the fluid stream to the solid particles. The fluid mechanics of fluidized beds is complicated enough without these additional problems of reaction and mass transfer, but fortunately heat transfer is rapid in fluidized beds, which are thus essentially isothermal.

For liquid-fluidized beds, which fluidize 'particulately' or 'smoothly', mass transfer rates can be related to those in packed beds. Experiments with beds of ion-exchange resin beads [**14**] showed that the equation

$$Sh = \frac{k_M d}{D} = \frac{0.81}{\varepsilon} Re^{1/2} Sc^{1/3} \tag{8.18}$$

could be used to describe results in both packed- and liquid-fluidized beds (see § 7.2). The range of variables covered was not very wide. The *measured* value of ε was used, but for design purposes one must be able to *estimate* ε, which varies with Re in a fluidized bed. One expression for ε which has been proposed [**15**] is

$$\varepsilon = (u/u_t)^{1/n}, \tag{8.19}$$

in which u_t is the terminal velocity of free fall of a single particle through stagnant fluid. The value of n varies somewhat with Re, being 4.65 for $Re < 0.2$, decreasing to 2.39 for $Re > 500$.

It is gas-fluidized beds, however, which are of prime interest, and these fluidize 'aggregatively' or in a bubbling manner. This poses severe problems for the designer and these can only be outlined.

It may first be remarked that experiments on fluidized beds might be examined on the following alternative assumptions:

(*a*) that the particles are 'well-mixed' in the bed by circulation produced by the fluid;

(*b*) that the rate of circulation of particles is slow enough to be neglected in reaction calculations (i.e. the particles resemble a packed bed);

(*c*) that the fluid passes through the bed in plug flow;

(*d*) that the fluid passes through without any back-mixing, though the whole fluid does not have the same forward velocity, as it would if there were plug flow;

(*e*) that the fluid in a fluidized bed is well-mixed due to rapid circulation of the solid, which is also well-mixed.

For beds of different geometry, and with various substances at different flow rates, any one of these assumptions may seem reasonable. Nevertheless the interpretation of a given set of experimental measurements will give widely different results according to the model chosen to represent the system.

For liquid-fluidized beds assumptions (*a*) and (*c*) may be reasonable (see [**14**]), but for bubbling gas-fluidized beds (*c*) is not very reasonable and (*d*) seems a better picture of the situation.

The approach favoured by Davidson and Harrison [**16**], and others, is to consider the fluid mechanics of a single bubble in a fluidized bed. This is shown to approximate to a 'spherical-cap' form, similar to that of a large air bubble rising through water. In a fluidized bed, however, the fluidizing phase can pass *through* a bubble out into the particulately-fluidized phase (in some cases circulating round and back into the wake at the rear of the bubble). The mass transfer between the bubble and the particulately-fluidized phase (within which the reaction takes place at the surface of the particles) and hence the conversion, can be calculated, as can the 'bypassing' loss due to fluid passing rapidly through the bed in bubbles without contacting the solid.

A major difficulty of this approach is that in a violently bubbling bed, such as is used industrially, the size, rise-velocity and general behaviour of the bubbles may not easily be estimated from experiments at low fluidizing velocities where the bed is comparatively quiet, and bubbles are well-separated and well-defined.

A second, related, approach favoured by many and described by Kunii and Levenspiel [**17**] is to regard a gas-fluidized bed as consisting of two regions. The first is the 'bubble phase', the second the 'emulsion phase'. In either region the flow pattern can be plug flow or perfectly mixed, while there is mass transfer between the two regions. Chemical reaction occurs predominantly in the emulsion phase.

These two approaches have gathered their own supporters, but attempts to apply either of them to industrial fluidized-bed reactors have not been very successful. Grace [**18**], in a review of the modelling of fluidized-bed reactors, points out that the experimental evidence does not support either representation very well.

There has recently been heightened interest in the subject of fluidization as a result of work in the fluidized combustion of coal, which opens up the possibility of the use of fluidization on a gigantic scale. In principle fluidized combustion enables coals of high ash content to be burnt, and also coals of high sulphur content. This is because the burning coal particles are but a small proportion of the fluidized bed. The remainder is of ash particles, to which can be added crushed limestone, which will react with the SO_2 produced by combustion of the sulphur in the coal. Thus SO_2 can be removed from the flue gases without the need for expensive scrubbing operations. For large fluidized-bed combustors Levenspiel [**19**] has recently proposed a new 'plume' model.

In these combustors the heat is removed by banks of steam-raising tubes which are submerged in the fluidized bed. In such situations the fluid mechanics are indeed affected by the presence of the tubes. Very

Fig. 45. The 'cat-cracker' at Pernis refinery. (A Shell photograph.)

high heat-transfer coefficients between the hot bed of ash (containing burning coal particles) and the tubes can be obtained.

Fluidization is a topic on its own in chemical engineering and in this book we have attempted only to point out the importance and design difficulties of fluidized-bed reactors. Some idea of the size of a 'cat-cracker' can be gauged from the picture shown in Fig. 45, where there are two separate fluidized beds, each fulfilling a distinctive role in the cracking process. In the 'reactor' the catalyst, finely-divided alumina, is contacted with a heated stream of oil vapour as is shown in Fig. 46. Here the cracking reaction results in carbon being deposited on the catalyst particles. These are transferred to the 'regenerator', in which the carbon is burnt off by the fluidizing stream of air. A cat-cracker in operation is a formidable sight – and sound!

Fig. 46. Schematic flow-diagram of a 'cat-cracker'.

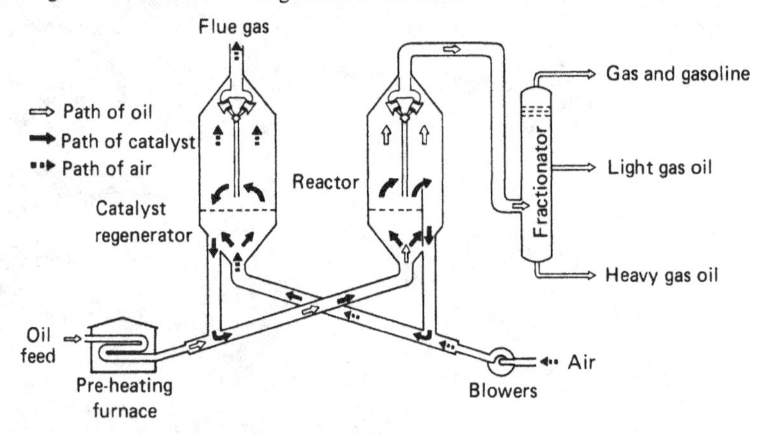

Example 8.11

The liquid-fluidized bed of resin spheres of Example 8.10 is part of an ion-exchange apparatus. Estimate the fluid–solid mass transfer coefficient, and *Sh*, in the bed when the fluidizing velocity is five times the minimum fluidizing velocity. For this situation the value of n in equation (8.19) is given by $n = 4{\cdot}45\ Re^{-0.1}$. The terminal velocity of the particles can be estimated from

$u_t = (\rho_s - \rho)d^2 g/18\mu.$

As in Example 8.10, $\rho_s = 1200\ \text{kg m}^{-3}$, $d = 0.9\ \text{mm}$, and $\mu = 10^{-3}\ \text{N s m}^{-2}$. The diffusion coefficient of the reagent in the fluidizing water is $1.5 \times 10^{-9}\ \text{m}^2\ \text{s}^{-1}$.

[*Answers.* $k_M = 6.3 \times 10^{-5}\ \text{m s}^{-1}$, $Sh = 38$.]

Example 8.12

A spherical gas bubble dissolves in a liquid which is involatile and contains no solute gas in the bulk. The mass transfer is liquid-side controlled, and the mass-transfer coefficient k_L ($m\,s^{-1}$) is in this case constant.

If the bubble contains solute gas only, show that the time for it to dissolve completely is given by

$$\left(\frac{3v_0}{4\pi}\right)^{1/3}\frac{H}{k_L RT},$$

where v_0 (m^3) is the initial volume, H ($N\,m\,kmol^{-1}$) is the Henry's Law constant, R ($J\,kmol^{-1}\,K^{-1}$) is the gas constant, and T (K) is the temperature.

Example 8.13

If the bubble in Example 8.12 contains an initial mole fraction X of insoluble gas, show that it contracts from volume v_0 to v_1 in a time t_1 given by

$$t_1=\frac{H}{(36\pi)^{1/3}k_L RT}\int_{v_1}^{v_0}\frac{dv}{v^{2/3}(1-Xv_0/v)}$$

Example 8.14

Oxygen in a stagnant gas is diffusing towards a burning carbon particle of radius R. The oxygen concentration far from the particle is c_b. Stating your assumptions clearly, show that the rate of consumption of oxygen by the particle is given by $4\pi D_G R c_b$, where D_G ($m^2\,s^{-1}$) is the diffusion coefficient of oxygen.

Example 8.15

Assuming the conditions in an atmospheric-pressure fluidized combustor are such that the results of Example 8.14 apply, calculate the burn-out time for 1 mm diameter carbon particles. The carbon density is 1400 $kg\,m^{-3}$, the oxygen mole fraction is 5%, the bed temperature is 900 °C, and $D_G=2\times10^{-4}\,m^2\,s^{-1}$.

[*Answer.* 142 s.]

Example 8.16

A fluidized combustor is fed continuously with 1 mm diameter carbon particles at the rate of 1 $kg\,s^{-1}$. Using the results of the previous example, calculate the weight of carbon in the bed.

[*Answer.* 57 kg.]

Example 8.17

Show that the result of Example 8.14 is that the Sherwood number is constant. Contrast this with the situation assumed in Example 8.12.

Example 8.18

The fluidized combustor of Example 8.16 is fed continuously with particles of radius R_0. Show that the number distribution of particles in the steady state is given by $P = kr$, where Pdr is the number of particles of radius between r and $r + dr$, and k is a constant.

At time $t = 0$, the feed of particles is turned off. If everything else stays the same, show that the largest particles remaining at time t will have a radius of R where

$$R^2 = R_0^2(1 - t/\tau)$$

and τ is the burn-out time for particles of radius R_0.

Example 8.19

Show that in Example 8.18, the generation of heat, Q, will fall off with time as

$$Q = Q_0(1 - t/\tau)^{3/2},$$

where Q_0 is the generation of heat when the feed of particles to the combustor is turned off.

Symbols

c Concentration, kmol m^{-3}.
d Equivalent diameter of packing piece, m.
D Diffusion coefficient, $m^2\,s^{-1}$.
E Enhancement, see §8.2.
g Acceleration of gravity, m s^{-2}.
Gr Modified Grashof number, $Gr = \mu^2/\rho^2 g d^3$.
k Reaction velocity constant.
k_G, k_L Gas- or liquid-film mass-transfer coefficient, m s^{-1}.
K_G, K_L Overall gas or liquid mass-transfer coefficient, m s^{-1}.
H 'Henry's Law Constant' – the ratio of gaseous and liquid concentrations at equilibrium.
m Reaction order.
M k δ^2/D, see equation (8.5).
N_A Flux of reagent, kmol $m^{-2}\,s^{-1}$.
$\dot{N}$ Moles destroyed per second per particle.
r Radius, m.
R Starting radius, or maximum radius, m.

Re Reynolds number, based on particle equivalent diameter and fluid superficial velocity, $Re = \rho ud/\mu$.
s Constant in equation (8.12).
Sc Schmidt number, $Sc = \mu/\rho D$.
Sh Sherwood number, e.g. $Sh_L = k_L d/D_L$.
t Time.
u Fluid superficial velocity, m s^{-1}.
x Co-ordinate distance, m.
δ Equivalent stagnant-film thickness, m.
ε Voidage of packed, or fluidized, bed.
μ Fluid viscosity, N s m^{-2}.
ρ Fluid density, kg m^{-3}.
ξ Fractional distance into stagnant film, x/δ.
ψ Concentration ratio, c/c_i, see equation (8.5).

References

1. van Landeghem, H., *Chem. Engng Sci.*, 1980, **35**, 1912.
2. Danckwerts, P. V. and Sharma, M. M., *The Chemical Engineer*, Inst. of Chem. Eng. London, Oct. 1966, p. CE 244.
3. Danckwerts, P. V., *Gas–Liquid Reactions*, (McGraw-Hill, New York, 1970).
4. Hatta, S., *Technol. Repts. Tohoku Imp. University*, 1928–9, **8**, 1.
5. Joosten, G. E. H., Maatman, H., Prins, W. and Stamhuis, E. J., *Chem. Engng Sci.*, 1980, **35**, 223.
6. 'Bubble Column Reactors' in *Chemical Reactors*, ed. Fogler, H. S., Amer. Chem. Soc. Symposium Series 168, (Amer. Chem. Soc., Washington, 1981).
7. Andrew, S. P. S., *Trans. Inst. Chem. Eng. London*, 1982, **60**, 3.
8. Smith, S. R. L., *Phil. Trans. Roy. Soc. Lond.*, 1980, **B290**, 341.
9. Bolton, D. H. and Ousby, J. C., *The ICI Deep Shaft Aeration Process for Effluent Treatment*, ICI Agricultural Division, Billingham, England, 1975.
10. Bryant, J., in *Advances in Biochemical Engineering*, Vol. 5, (Springer-Verlag, Berlin, 1977).
11. Billmeyer, F. W. Jr, *Textbook of Polymer Science* (John Wiley, New York, 1962).
12. Yagi, S. and Kunii, D. *5th Int. Symposium on Combustion* (Reinhold, New York, 1955), p. 231.
13. Cornish, A. R. H., *Trans. Inst. Chem. Eng.*, 1965, **43**, 321.
14. Snowdon, C. B. and Turner, J. C. R., *Proceedings of the International Symposium on Fluidisation*, ed. A. A. H. Drinkenburg (Netherlands University Press, Amsterdam, 1967).
15. Richardson, J. F. and Zaki, W. N., *Trans. Inst. Chem. Eng.*, 1954, **32**, 35.
16. Davidson, J. F. and Harrison, D., *Fluidised Particles* (Cambridge University Press, 1963).
17. Kunii, D. and Levenspiel, O., *Fluidization Engineering* (John Wiley, New York, 1969).
18. Grace, J. R., *Chemical Reactors*, Amer. Chem. Soc. Symposium Series 168, ed. H. Scott Fogler (Amer. Chem. Soc., Washington, D.C., 1981) p. 3.
19. Park, D., Levenspiel, O. and Fitzgerald, T. J., *Chem. Engng Sci.*, 1980, **35**, 295.

9

Some further temperature effects in reactors

9.1 Introduction

Earlier chapters have included the effect of temperature on the kinetics of chemical reaction, and the major effects that temperature gradients can have in tubular reactors. The appendices to Chapter 3 show the vast differences in design volume which can arise when a tubular reactor is operated adiabatically instead of isothermally. Again, in Chapter 7, we saw that temperature effects in catalyst particles can lead to radically different behaviour, even including multiple steady states, from that expected under isothermal conditions. Nevertheless, in much of what has gone before we have assumed that the temperature of the reactor is given. The design then proceeded from a knowledge of the material flows and of the kinetics at that particular temperature. In general, however, the operating temperature is not known in advance, but has to be calculated. It may be that controlled variations of temperature within the reactor can be of great value.

The effect of temperature upon chemical reaction rate can be so marked that it determines the nature of the reactor, in the sense that the selectivity of the reactor, i.e. *what* it makes (as opposed to how much), depends crucially upon the temperature. As well as this, the stability of the reactor, its controllability, may be primarily determined by temperature effects.

In recent years there has been a rapid growth of interest in this area, which in some ways mirrors the earlier (and continuing) interest of physical chemists in chain reactions in such areas as polymerization kinetics and explosion kinetics. Indeed there has been so much mathematical theory that it has tended to cause industrial practitioners to lose interest. A recent paper by Pismen [**1**] gives some 300 references on kinetic instabilities in reactors (including living 'reactors'). This, introductory, text will not be able to go far along this path although it is one being followed with enthusiasm in graduate schools around the world.

The most interesting thermal effects occur when the reaction is exothermic. When an exothermic reaction is carried out adiabatically and batchwise the temperature will rise as the reaction proceeds until the reaction, and the generation of heat, cease. The rate of reaction, and hence the generation of heat, will at first increase as the temperature rises. The temperature rise is related to the extent of reaction, and the higher this is the lower will be the concentrations of the remaining reagents. Thus eventually the reaction slows down owing to depletion of the reagents, finally stopping altogether when they are totally consumed, as is shown in Fig. 47.

If the reaction is significantly reversible, then increase of temperature will cause a decrease in the equilibrium constant, and hence there will be a decrease in the maximum possible conversion as the temperature rises. This effect will cause the reaction to cease, and the temperature to level off, at lower conversions than in the case where the reaction is effectively irreversible.

In the case of a continuous reactor these considerations are modified by the flow of reagents and products. It is necessary to carry out a material *and* thermal balance on an element of the system to discern its behaviour. In principle this procedure is no more than an extension of the treatment for the material balance outlined first in Chapter 2, but the balance equations must allow for heat transfer to or from the system

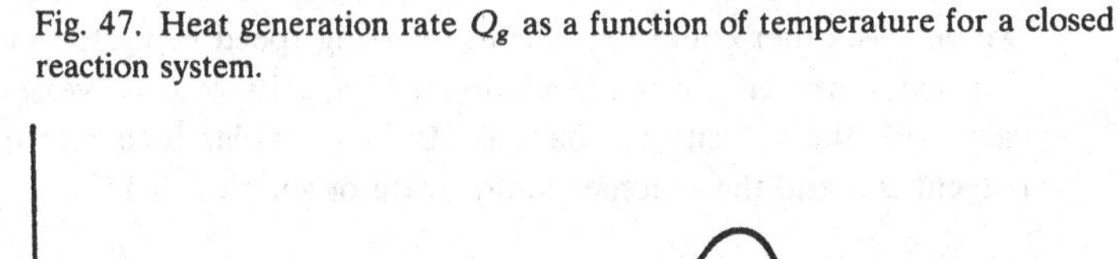

Fig. 47. Heat generation rate Q_g as a function of temperature for a closed reaction system.

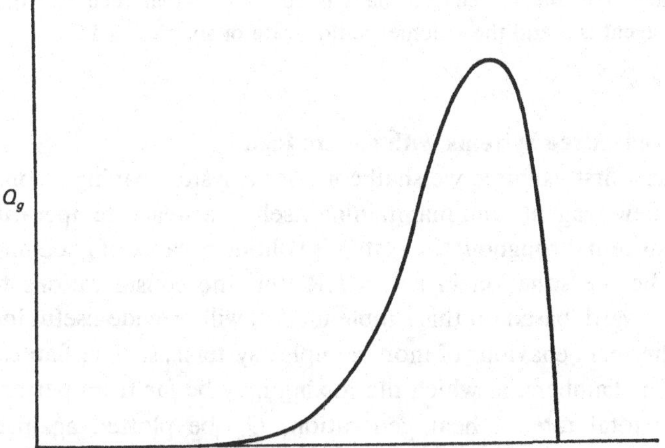

since it will not usually be adiabatic. Such heat exchange can radically alter the behaviour of the system, as will be shown.

***Example* 9.1**

In a continuous process an aqueous solution of an ester of a monobasic organic acid is hydrolysed with caustic soda solution in a C.S.T.R. which holds 6 m^3 liquid. Within this vessel there is a submerged cooling coil, which maintains the reaction temperature at 25 °C. Use the following data to estimate the necessary area of heat-transfer surface for conditions such that cooling water enters the coil at 15 °C and leaves at 20 °C. Neglect heat loss from the sides of the vessel.

Concentration, temperature and flow rate of ester solution: 1.0 kmol m^{-3}, 25 °C, 0.025 $m^3\,s^{-1}$, respectively.

Concentration, temperature and flow rate of alkali solution: 5.0 kmol m^{-3}, 20 °C, 0.010 $m^3\,s^{-1}$, respectively.

Reaction velocity constant at 25 °C: 0.11 $m^3\,kmol^{-1}\,s^{-1}$.

Heat of reaction: 1.46×10^7 J $kmol^{-1}$.

Heat transfer coefficient under the given conditions: 2280 J $m^{-2}\,s^{-1}\,K^{-1}$.

[*Answer.* 8.0 m^2.]

***Example* 9.2**

A second-order liquid phase reaction between reagents A and B is carried out in two continuous stirred tank reactors in series, each of equal capacity. Each vessel has a cooling coil; estimate the ratio of the areas of these coils needed to obtain equality of temperatures in the two vessels, other conditions such as stirring speed being the same.

In self-consistent units the volumetric capacity of each vessel has a value 10^4, the velocity constant is 10^{-2}, the molar feed rate of each reagent is 5 and the volumetric flow rate of solution is 12.

[*Answer.* 3.2 : 1.]

9.2 Well-mixed systems with steady feed

As a first example we shall consider a system having a constant supply of new reagents and maintaining itself at a steady temperature T which is uniform throughout the system's volume because of good mixing. This will be the situation in a C.S.T.R. but the considerations to be brought forward, based on this simple model, will provide useful insight into the thermal behaviour of more complex systems, such as flames and combustion chambers, in which the mixing may be far from perfect.

Let the total rate of heat generation, Q_g, be plotted against the temperature T, for fixed values of the reagent feed rates to the reactor.

The curve will be shown to have the sigmoid shape of Fig. 48, due to the opposing effects of increased velocity constant and diminished concentrations of reagents. At a high enough value of T the velocity constant is so great that virtually no unreacted reagent remains in the outflow from the system (the feed rates remaining the same) and thus a still higher value of T cannot make Q_g significantly greater, with the consequence that the curve must flatten off.

The equation of the curve is readily obtained if suitable assumptions are made. Let it be supposed for simplicity that the reaction is irreversible and that it is also first order, its speed being proportional to the concentration of one substance only.

Assuming perfect mixing and no density change we have (see Chapter 4)

$$c = c_0/(1 + kV/v), \tag{9.1}$$

where c is the concentration of the reference substance within, and leaving, the reactor of volume V, and feed is supplied at volume flow rate v and concentration c_0. We assume for the moment that the reactor is in a steady state. The rate at which heat is generated is

$$Q_g = -kVc\Delta H. \tag{9.2}$$

Now $k = Ze^{-E/RT}$, where T is the temperature prevailing in the C.S.T.R. Substituting this and equation (9.1) in equation (9.2) we obtain

$$Q_g = \frac{-vc_0\Delta H}{1 + \dfrac{v}{VZ} e^{E/RT}}. \tag{9.3}$$

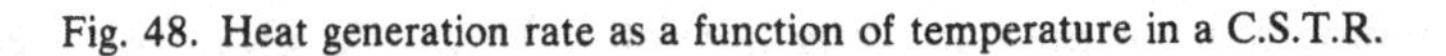

Fig. 48. Heat generation rate as a function of temperature in a C.S.T.R.

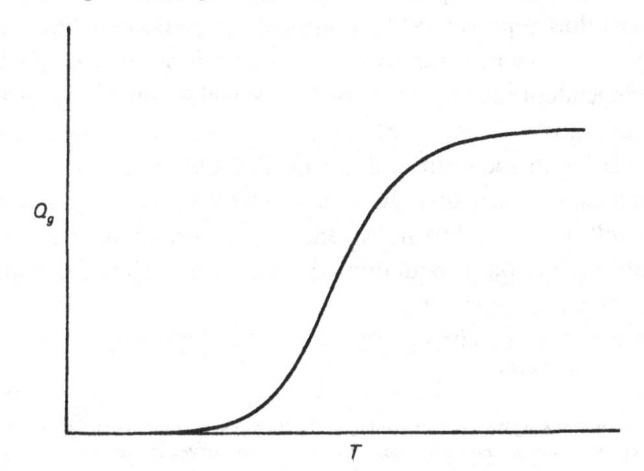

If Q_g is plotted as a function of T for fixed values of the other variables the curve will be found to be as in Fig. 48.

Considering now the rate Q_r at which heat is removed from the system, let it first be assumed that the system operates adiabatically, i.e. without passage of heat through its walls. In this case Q_r is simply the rate at which heat is carried away from the C.S.T.R. by the outgoing fluid. Therefore

$$Q_r = v\rho C(T - T_0), \tag{9.4}$$

where T_0 is the feed temperature and T, as before, is the temperature prevailing inside the vessel. C and ρ are the specific heat and density of the fluid respectively and it has been assumed, for simplicity, that these quantities have unchanged values as between inflow and outflow.

In the case of non-adiabatic operation a further term must be added to equation (9.4) and, under certain simplifying conditions, this will be of the form $UA(T - T_c)$. Here U and A refer to a heat-transfer coefficient and to a heat-transfer area respectively whilst T_c is the temperature, assumed to remain approximately unchanged, of the heating or cooling fluid. Hence the total of heat removal is

$$Q_r = v\rho C(T - T_0) + UA(T - T_c) \tag{9.5}$$

$$= T(v\rho C + UA) - v\rho C T_0 - UAT_c. \tag{9.6}$$

It will be seen that equations (9.4) and (9.6) are both linear functions of T and thus, for fixed values of the other variables, are represented by straight lines of gradients $v\rho C$ or $(v\rho C + UA)$ respectively.* A number of such lines are shown in Fig. 49, which also shows the heat generation curve of Fig. 48.

If a steady state is to occur in the system (i.e. a condition such that its temperature is neither rising nor falling), Q_g and Q_r must be equal. Steady states are thus represented by points of intersection of the heat generation curve with the heat removal line. The positions of such points are therefore dependent on the values of the variables which appear on the r.h.s. of equations (9.3) and (9.5), i.e. the variables which determine the positions relative to each other of the Q_g and Q_r curves on Fig. 49.

If the conditions are such that there is an intersection such as *a* on the figure, this will correspond to an extremely low, perhaps almost zero, rate of reaction. Inspection of equations (9.3) and (9.5) shows that this type of steady state is favoured by:

(*a*) low value of the velocity constant;
(*b*) low residence-time;

* If radiation makes a significant contribution to the heat loss from the system, Q_r is no longer represented by a straight line but rather by a curve convex to the T axis.

(*c*) small value of the heat of reaction;
(*d*) small feed rate vc_0;
(*e*) small values of T_0 or T_c;
(*f*) large values of U or A.

On the other hand, an intersection point such as *b* corresponds to almost complete reaction and this is favoured by the converse conditions. If such an intersection is obtained under conditions where no heat is being put into the system (i.e. when the heat transfer term in equation (9.5) is zero or positive) the reaction is said to be capable of being operated *autothermally.*

It will be seen that, on account of the sigmoid shape of the heat generation curve, intersections at either low or high degrees of conversion are more likely to occur than intersections at intermediate values. This is in general agreement with experience. Thus in the combustion of fuels the tendency is either for the temperature to be so low that the fuel is virtually not reacting (as in a pile of coal in the stockyard) or for it to be so high that the oxygen content of the combustion gases is reduced almost to zero.

Let it be supposed, however, that the relative positions of the sigmoid curve and the heat removal line are such that they intersect at three points, *c, d* and *e.* The first and last of these are of the same type as *a* and *b* respectively and call for no further comment. (Except perhaps to remark that if it is desired to operate at the upper steady state the lower one has to be bypassed, e.g. by a temporary period of heating.)

Point *d*, on the other hand, has unusual features: at this point, although heat generation and heat removal are equal, the system could not operate

Fig. 49. Heat generation and heat removal curves for a C.S.T.R.

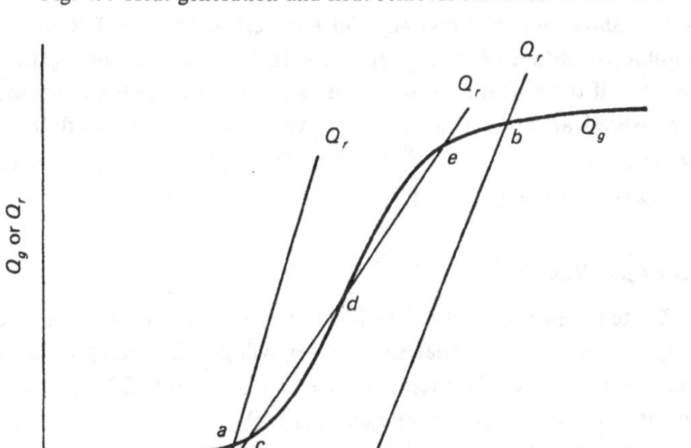

in a stable manner. Any slight upward fluctuation of temperature would make the heat generation greater than the heat removal and thus the temperature would continue to rise until point *e* was reached. Similarly, any slight downward fluctuation of temperature would result in a further fall of temperature until point *c* was attained. Therefore *d*, although it corresponds to a stationary state of the system, does not correspond to a *stable* stationary state. The question of the stability of the three states *c*, *d* and *e* will be discussed further in the next section.

Example 9.3

A reaction is being carried out in a C.S.T.R. of volume $10\ m^3$. The feed solution contains a reagent at concentration $5\ kmol\ m^{-3}$ and is supplied at a rate of $10^{-2}\ m^3\ s^{-1}$. The reagent decomposes irreversibly according to a first-order reaction, liberating 2×10^7 J of heat per kmol reacting. The velocity constant of the reaction is given by
$k = 10^{13}\ e^{-12000/T}$.

The density of the solution is $850\ kg\ m^{-3}$ and can be taken as being independent of temperature or extent of reaction. The specific heat of the solution is similarly constant and equals $2200\ J\ kg^{-1}$.

By the method illustrated in Fig. 49, calculate the reaction temperature and extent of reaction if the feed solution is provided at (*a*) 290 K, (*b*) 300 K, (*c*) 310 K.

[*Answers.* (*a*) 290.5 K, about 1%; (*b*) either 303 K, 6% or 349 K, 92%; (*c*) 362 K, 98%.]

Example 9.4

With the reactor of Example 9.3 it is found that the occurrence of a side reaction requires that the reactor temperature must remain below 340 K. Show that if a cooling coil is fitted in the C.S.T.R. it will be possible to obtain 80% conversion with no problems of 'ignition' at start-up. If the coolant temperature, T_c, is 310 K, and the feed solution is provided at 310 K, calculate the value of UA required from the cooling coil.

[*Answer.* $UA = 9000\ J\ s^{-1}\ K^{-1}$.]

Example 9.5

The temperature of the solution fed to the reactor of Example 9.4 drops to 300 K. Show that the reactor will continue to produce 80% conversion if the coolant temperature, T_c, is raised to 331 K. Comment on the role now played by the cooling coil.

One of the most important of the variables affecting the relative positions of the Q_g and Q_r curves is the flow rate. From equation (9.4) it is seen that an increase in v increases the slope of the Q_r line. From equation (9.3) we see that the shape of the Q_g curve is also changed, but the intersection of the Q_r line with it will move to the left. It could thus occur that, in place of an intersection of type *b*, the only intersection becomes one of type *a*. This would correspond to a 'blowing out' of the reaction as occurs sometimes in combustion when the inflow of the cold gas is too great relative to the amount of heat evolved. This is to say autothermic reaction would no longer be possible at a high degree of conversion.

Conversely if v were to be greatly reduced, it could again occur that autothermic reaction would no longer be possible. At very low feed rates the inevitable heat losses from the system (these heat losses being equivalent to the presence of the second term in equation (9.5), even though no cooling were in use) would become relatively far more important and would result in the Q_r curve being moved to the left more than the Q_g line. Thus the *b* type of intersection might no longer be obtained. An example of this type of behaviour is the quenching of combustion in a stove when the air supply is reduced too much. Thus, it can occur that an autothermic state of reaction is possible only between lower and upper limits of flow [**2**, **3**]. In general, however, these limits are far apart and between them conversion is usually fairly complete.

Example 9.6

Show that in the reactor of Example 9.3 it will no longer be possible to obtain a high level of conversion if the feed rate is doubled, its temperature being 300 K. If, at this feed rate of $2\times10^{-2}\ m^3\ s^{-1}$, the feed temperature is raised to 310 K, show that a conversion of over 90% is obtainable, without the need for any heating at start-up.

Example 9.7

Following damage to the lagging, heat losses during the winter from the reactor in Example 9.3 are found to be given by an expression of the form

heat losses $= 5000(T-280)\ J\ s^{-1}$.

Show that the reaction will be substantially 'quenched' if the feed rate is dropped to $2\times10^{-3}\ m^3\ s^{-1}$, its temperature remaining at 310 K. What conversion will now be obtained?

[*Answer.* 12% (compare 98% for Example 9.3(*c*).]

So far the discussion has been concerned with simple irreversible reactions. For other types of reaction the heat generation curves may present more elaborate features and there may occur a larger number of intersections with the heat removal line and thus a larger number of stationary states.

Fig. 50 shows the heat generation curve for a simple reversible exothermic reaction when it occurs in a single-stage C.S.T.R. For such reactions the maximum attainable degree of conversion diminishes with rise of temperature. In kinetic terms this is because the net speed of reaction begins to diminish above a certain temperature, the backward reaction being accelerated more than the forward reaction. This accounts for the diminishing rate of heat generation on the right of Fig. 50 and at a high enough temperature this rate approaches zero (or it could become negative if reaction product were already present in the feed to the vessel). It follows that a heat removal curve such as Q_r (1), which results in a steady state at point *a* gives rise to a lower attained degree of conversion than a heat removal curve such as Q_r (2), which corresponds to more vigorous cooling. The best conditions with regard to conversion would evidently be attained by achieving a heat removal curve which strikes the heat generation curve near to its peak. However, if the cooling

Fig. 50. Heat generation and heat removal in a C.S.T.R. Example of a reversible reaction.

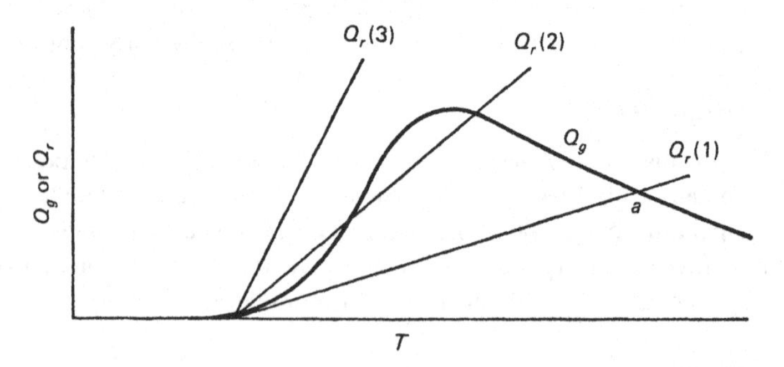

Fig. 51. Heat generation curve for two consecutive reactions, both exothermic. (After Bilous and Amundson, *A.I.Ch.E. Journal*, 1955, **1**, 513.)

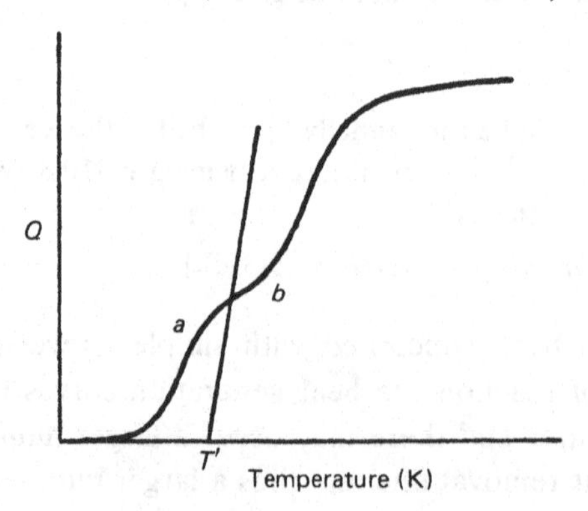

rate were made only slightly greater, corresponding to a line in the region of Q_r (3), this could result in reaction dying out. To obtain optimum cooling may therefore require very sensitive control.

Figs. 51, 52 and 53 refer to consecutive reactions of the type

$$A \xrightarrow{\Delta H_1} B \xrightarrow{\Delta H_2} C,$$

for different signs of the heats of reaction and for suitably chosen values of the activation energies and frequency factors. (If the second reaction is much faster than the first, the distinctive parts of these curves tend to merge into each other, resulting in a curve more similar to that of Fig. 48). The existence of these more complicated types of heat generation curve and the mathematical treatment of the conditions of stability have both been demonstrated [**4**].

Fig. 52. As for Fig. 51, but second reaction endothermic.

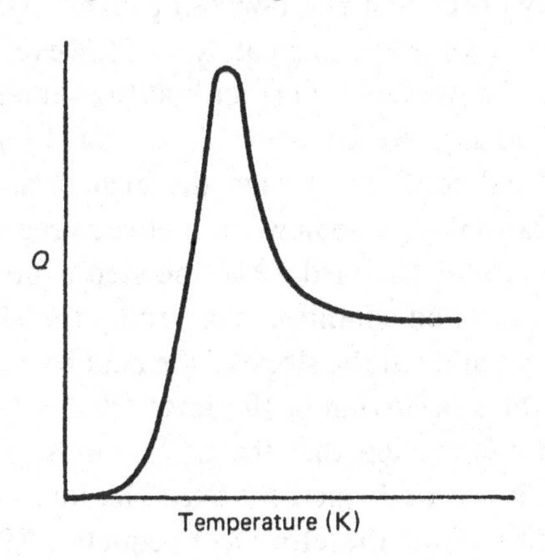

Fig. 53. As for Fig. 51, but first reaction endothermic.

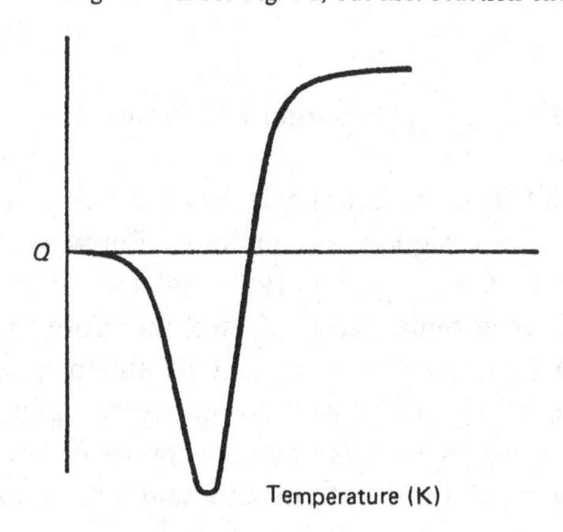

It will be seen that the situation represented in Fig. 51, which refers to the case where both of the consecutive reactions are exothermic, allows of the possibility of five steady states, and three of them can be stable, as can happen in ethylene polymerization [**5**]. This situation is actually of great practical importance and especially when it is substance B that is the desired product whilst C is not desired. This commonly occurs in oxidation reactions; in these the desired product may frequently oxidize further to give wasteful end-products by highly exothermic reactions. These tend to raise the temperature to such an extent that B is almost entirely consumed. A good example of this is the production of phthalic anhydride by oxidation of naphthalene. Unless the temperature is carefully controlled the phthalic anhydride may oxidize further to CO_2 and H_2O and the reaction is therefore of the type under discussion.

In such instances the possibility of obtaining a useful yield of B most obviously depends on the velocity of the first reaction being much higher than that of the second (e.g. by use of a suitable catalyst). However, it also depends on the possibility of maintaining the reactor at the required steady state. The central 'step' in Fig. 51 corresponds to the desired product and therefore the essential condition is that the heat removal line shall intersect the heat generation curve somewhere between regions *a* and *b*. This condition is more readily satisfied if (*a*) the step is broad (which depends on the above-mentioned condition concerning the relative velocities of the two reactions), and (*b*) the slope of the heat removal line, as determined by the variables occurring in the term $(v\rho C + UA)$ in equation (9.6), is large. Also it is evident that the Q_r line must start from some appropriately chosen intercept T' on the T axis. This intercept is determined by the condition $Q_r = 0$ and therefore from equation (9.6) by the relation

$$T' = \frac{v\rho C T_0 + UAT_c}{v\rho C + UA}. \tag{9.7}$$

The variables occurring on the r.h.s. of equation (9.7) must therefore be correctly chosen.

Because of the stringency of these conditions, it may not be possible [**6**] to achieve stable operation under adiabatic conditions. For when the term UA is zero it will be seen from equation (9.7) that the intercept T' is necessarily equal to the feed temperature T_0 and therefore there may be insufficient freedom for manoeuvre. In fact to obtain a heat removal line intersecting the heat generation curve between regions *a* and *b* on Fig. 51 it may be necessary to make an appropriate choice of the magnitudes of both the heat-transfer term UA and the coolant temperature T_c. Alternatively the 'degrees of freedom' in the problem can be increased by shifting the position of the heat generation curve itself. Therefore it may occur that the feed concentration of the reagents has to lie between certain limits.

This section has so far considered how temperature effects can lead to alternative steady states of a C.S.T.R. The next section will consider the stability of such steady states. If an *autocatalytic* reaction is carried out in a C.S.T.R., it is possible to obtain behaviour, under isothermal conditions, which mirrors these temperature effects. Gray and Scott [**16**] consider a number of such situations. For example an autocatalytic reaction can show behaviour similar to the adiabatic operation of an exothermic reaction.

9.3 The stability of a C.S.T.R.

In the previous section we saw how it might be possible for a C.S.T.R. to operate in several different steady states, given the same conditions of feed and cooling. We shall now examine the question of the stability of these states in a little more detail. For a fuller exposition than is possible here, references [**1–6**] can be mentioned again. There is a good introduction to the topic by Aris [**7**], and also a text by Perlmutter entitled *Stability of Chemical Reactors* [**8**].

In the previous section we saw that the middle intersection, *d*, of Fig. 49 represented an unstable state, in that slight deviations from the point *d* caused the system to move to either *c* or *e*, the two stable states.

It was there stated that for a state to be stable the heat removal curve should have a greater slope than the heat generation curve at the point of intersection. this is indeed a *necessary* condition, but it is not sufficient. When using Fig. 49 to examine the stability of a state, we are implicitly assuming that any perturbation in the temperature (δT) will lead to that perturbation in the reaction rate (and hence heat generation) which is shown by the curve in Fig. 49. For a state to be stable it must be shown that the system returns to that state for *any* (small) perturbations in temperature and concentration which may occur, not only for perturbations in temperature coupled with those changes in concentration related to them by the steady-state curve.

The method is to linearize the response of the system to small perturbations; two first-order differential equations are obtained (for the mass balance on a reference component, and the heat balance on the reactor). These are coupled, but can easily be uncoupled, leading to two *second-order* differential equations, for each of which the solution is of the form, for example, $\delta T = A\,e^{m_1 t} + B\,e^{m_2 t}$, where in general m_1 and m_2 may be complex numbers. If the perturbation, T, is to die away, both m_1 and m_2 must have negative real parts. Hence there are *two* criteria which must be satisfied for stability. One of them turns out to be the same as our 'slopes' criterion. The other does not have such an obvious physical meaning.

For the adiabatic C.S.T.R., the second criterion is actually less demanding than the first, so in fact the 'slopes' criterion is both necessary *and* sufficient.

For the non-adiabatic C.S.T.R., any states such as d are indeed unstable, but it can arise that a state apparently stable (on the 'slopes' criterion) is actually unstable. This will depend upon the method of heat transfer, which affects the values of m_1 and m_2, and may result in one, or both, of them having positive real parts.

As well as determining whether a desired state is stable to small perturbations, we should also consider how to *get to* this state (e.g. at start-up) from conditions *a long way* from those at the desired state. This will be done in the following section.

9.4 The start-up of a C.S.T.R.

Where there is only one stable state of a C.S.T.R., for a given feed and heat transfer, then it will eventually reach that state, whatever conditions of temperature and concentration obtain inside the reactor at time $t=0$, The concentrations inside at $t=0$ may not be 'compatible' with the feed, in that they cannot be obtained by reaction from the feed composition. But this does not matter; the incompatible fluid will be 'washed out' of the system.

The approach to the steady state may be by a somewhat roundabout path; for example, the temperature may initially move in the 'wrong' direction, before returning to end up at the steady-state value.

Where the system has two possible steady states, it is not a simple matter to determine which one the system will move towards from a given starting condition. If we choose starting conditions of concentration which are compatible with the feed, then we can plot paths on a diagram of extent of reaction versus temperature. These paths can be obtained by numerical computation of the course of reaction from the given starting conditions, using the non-steady-state differential equations of composition and temperature.

Bilous and Amundson [**4**] gave some striking examples, and many similar diagrams have appeared in the subsequent literature. Behaviour such as is shown in Fig. 54 may result. The diagram is divided into two regions by a line (shown dashed) passing through the unstable intersection *d*. If the reactor starts from any condition to the left of the line, it will move to stable state *c* (compare Fig. 49), whereas it will move to state *e* from any starting point to the right of the line.

Three points should be noted: firstly, the positions of *c*, *d* and *e* are those determined by the steady-state conditions of feed and heat transfer, as in Fig. 49. Secondly, the approach to *c* or *e* may be somewhat indirect! Thirdly, two starting conditions very close together (but on opposite sides of the dashed line) lead to very different final reactor conditions, and this can happen, for instance, if the temperatures of both starting conditions are above (or below) that of the unstable intersection, *d*.

Fig. 54. Approach paths to the steady state of a C.S.T.R.

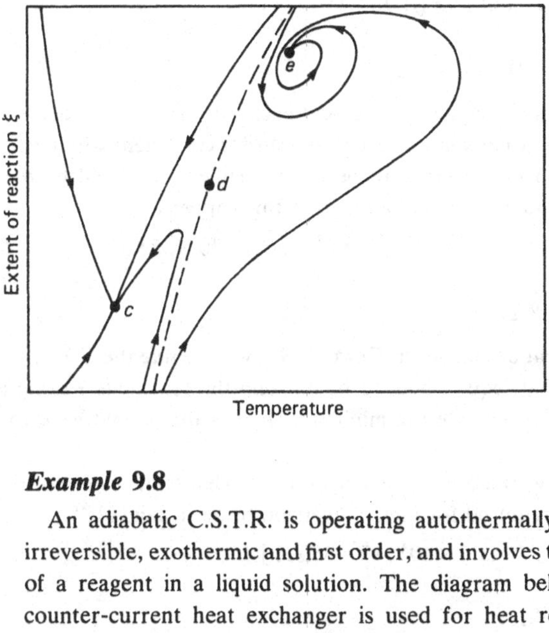

***Example* 9.8**

An adiabatic C.S.T.R. is operating autothermally. The reaction is irreversible, exothermic and first order, and involves the decomposition of a reagent in a liquid solution. The diagram below shows how a counter-current heat exchanger is used for heat recovery from the product stream. The temperature difference in the heat exchanger is constant.

Use the data below to calculate the temperature and the conversion in the reactor, which is operating in its higher-temperature operating state. Discuss how such a reactor system should be started up.

It will be helpful to plot $T_3 - T_0$ and $T_2 - T_1$ as functions of the reactor temperature T_2.

Feed concentration of reagent 4.4 kmol m^{-3}
Feed flow rate 10^{-4} m^3 s^{-1}
Reactor volume 1.5×10^{-3} m^3
Feed temperature $T_0 = 300$ K

Heat exchanger area = 0.4 m^2
Heat-transfer coefficient = 550 W m^{-2} K^{-1}
Heat capacity of solution = 4×10^6 J m^{-3} K^{-1}
Heat of reaction = 1.76×10^8 J kmol^{-1}

Velocity constant k, s^{-1}	0.025	0.061	0.16	0.47	1.26	67
Temperature T, K	440	460	490	520	555	870

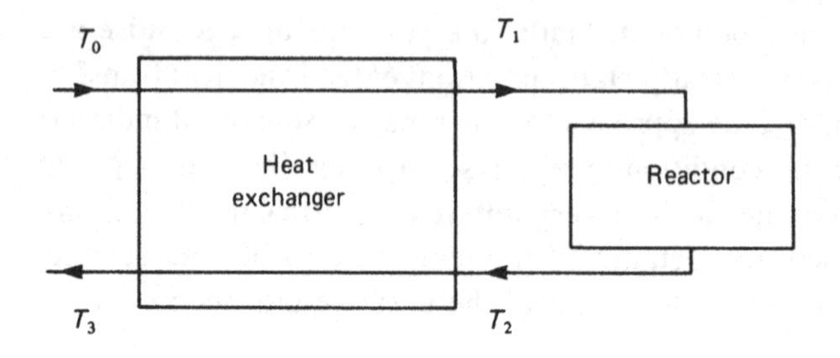

[*Answers.* About 595 K, 98%].

Example 9.9

Discuss what happens if the heat exchanger in the previous example fouls up. At what value of the heat-transfer coefficient will the reaction at high conversion cease to be maintainable? What will be the temperature and conversion just before this happens?

[*Answers.* About 320 W m^{-2} K^{-1}, about 520 K, 85%.]

Example 9.10

If the heat exchanger in Example 9.8 were *inside* the C.S.T.R., show that it would require heat to be put *into* the reactor to sustain a high conversion. Calculate the minimum amount that would have to be so provided.

If there were no heat exchanger at all, what temperature of the feed would be required for high conversion to be sustainable?

[*Answers.* About 2×10^4 J s^{-1}. 350 K.]

Example 9.11

An autocatalytic reaction $A \rightarrow B$ is carried out in solution in a C.S.T.R. at a fixed temperature. The reaction rate expression is of the form $-r_A = kc_Ac_B$.

The feed to the C.S.T.R. contains concentrations c_A^0 and c_B^0. Show that the fractional conversion of A to B, x, is given, in the steady state, by

$x = \theta(1-x)(R+x)$,

where $\theta = kVc_A^0/v$ and $R = c_B^0/c_A^0$.

Show that the maximum production of B will occur when $\theta = 2(1-R)/(1+R)^2$.

Example 9.12

If in the previous example $R = 0$, discuss the start-up of the reactor. What is the minimum value of θ for which any steady-state decomposition of A in the reactor is possible? Describe what happens if θ is reduced below this value.

[*Answer.* $\theta = 1$.]

Example 9.13

For the reactor of Example 9.11, write down the non-steady-state equations for c_A and c_B. By considering small perturbations, p_A and p_B, of c_A and c_B about their steady-state values, derive the differential equations for p_A and p_B. Investigate the stability of the steady state (by considering the solutions of these equations, and whether the perturbations will die away with time).

Example 9.14

Consider the non-autocatalytic decomposition of A in an *adiabatic* C.S.T.R. Write down the steady-state equations for c_A and T, the temperature of the reactor. By considering small perturbations, p_A and p_T, of these variables about their steady-state values derive the differential equation for p_A and p_T. Draw the analogy between these equations and those derived in the previous example. Discuss the stability of this reactor.

9.5 Limit cycles and oscillating reactions

In § 9.3 we considered perturbations about a steady-state solution to determine the stability of that solution. Under certain conditions of heat transfer it can occur that stable intersections, similar to points *c* and *e* of Fig. 49, are no longer obtained, but that the diagram is divided into two regions by a 'limit cycle', as in Fig. 55 [**9**]. This limit cycle encloses a point of unstable equilibrium, analogous to intersection *d* of Fig. 49. Any reactor starting at conditions either outside or inside the limit cycle will move towards the path defined by this cycle.

The progress of the reactor with time as it passes around the limit cycle is also shown in Fig. 55. The speed around the cycle is not uniform.

These curves have been calculated for model systems. An interesting paper by Fortuin and co-workers [**10**] gives practical results from a cooled C.S.T.R. within which the acid-catalysed liquid-phase reaction

```
                               H⁺
CH₂—CH —CH₂OH + H₂O ──▶ HOCH₂ . CHOH . CH₂OH
  \  /
   O
```

was carried out. This mildly exothermic reaction could be controlled in a limit cycle of a period of a few minutes. The amplitude of the temperature oscillations was some ±5 K, but concentration fluctuations of some ±30% were obtained. The agreement between the calculations and the measurements was good, and was much improved when the heat capacity of the solid parts of the reactor (tank, stirrer, etc.) was taken into consideration.

Fig. 55. Limit cycle in a two-phase reactor. The mole fraction of a given reagent in one of the phases is plotted against the reactor temperature (for several paths, upper graph) and time (for one path, lower graph). The limit cycle is the heavy closed loop, *S* is the point of unstable equilibrium (after Schmitz and Amundson, *Chem. Engng. Sci.*, 1963, **18**, 265).

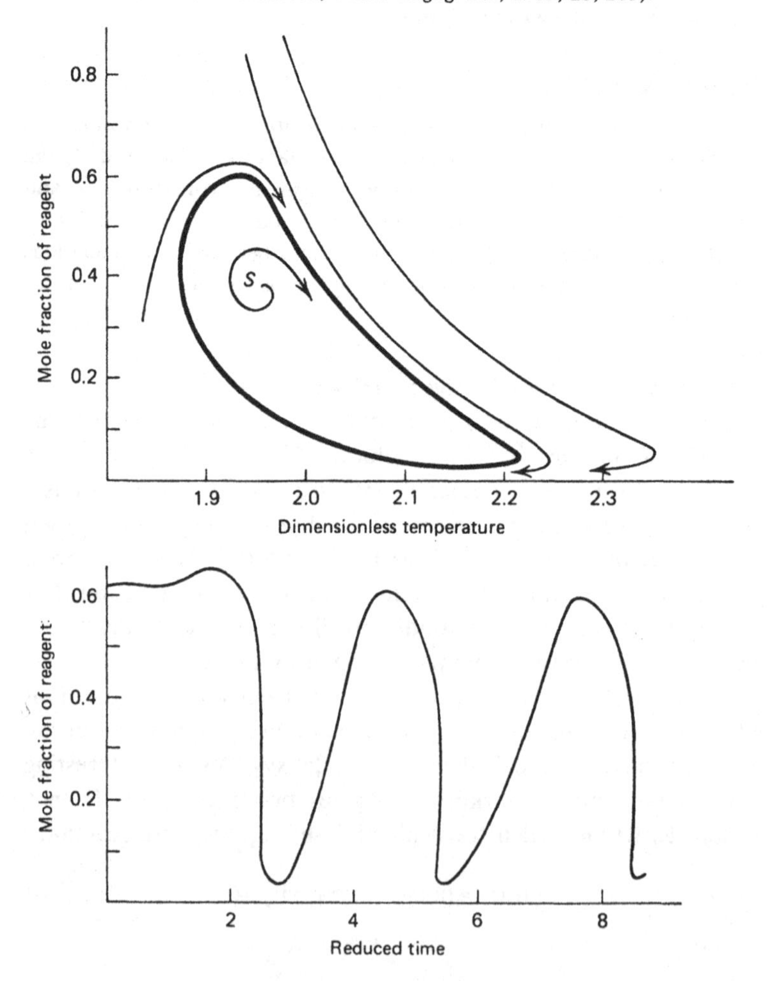

Similar sustained oscillations can also occur in combustion kinetics, see [**11**], and were investigated in the case of the oxidation of ether with air, which can show the 'cool-flame' phenomenon [**12**]. Fig. 56 shows the temperature fluctuations arising.

This type of behaviour has recently become more widely recognised; it can occur, for example, in many catalytic systems, see, for example, [**13**]. As Pismen [**1**] says, nobody can tell how many such processes may be found in the waste-basket of chemical and engineering research, while in biological phenomena oscillatory solutions are often positively preferred.

Nevertheless, it would be fair to say that bounded oscillatory behaviour is still unusual in reactor performance. Non-oscillatory instability is a much more likely danger for the plant designer and operator to have to contend with.

9.6 The plug-flow reactor. Parametric sensitivity

In the last sections we have considered the behaviour, and stability, of a C.S.T.R. with or without heat transfer. Let us now consider a plug-flow reactor from the same point of view; once again the most interesting effects arise with an exothermic reaction.

The adiabatic plug-flow reactor is always stable, in the strict sense that small perturbations in, for example, concentration or temperature are 'carried downstream and out of the reactor'. Again, a strictly plug-flow

Fig. 56. Regular temperature fluctuations associated with an oscillating reaction in a C.S.T.R. Two runs at different 'base' temperature show the marked temperature-dependence of the frequency of the oscillations (after Dutton, Ph.D. Thesis, London University, March 1968).

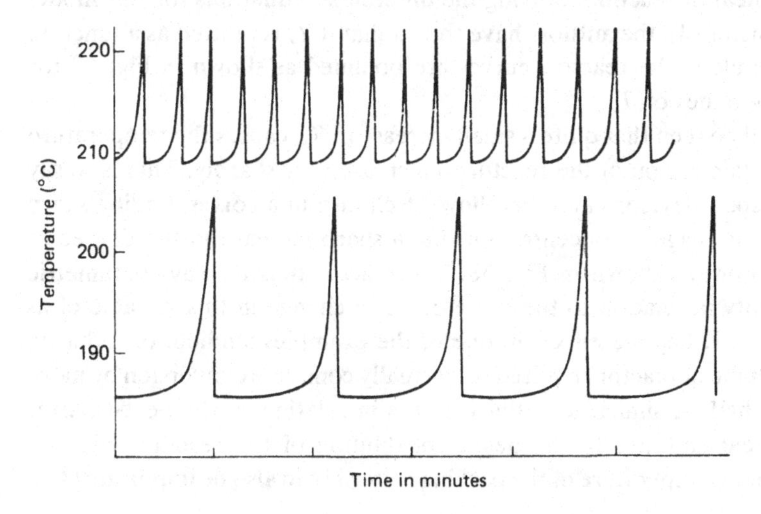

adiabatic reactor has only one possible steady state for given feed conditions. This may be deduced from equation (3.3),

$$V_r = G \int_{y_{in}}^{y_{out}} \frac{dy}{r}, \tag{9.8}$$

which shows that the volume of the reactor, V_r, is a monotonic function of the conversion, y, since the reaction rate r is always positive. Thus a given value of y_{out} will be associated with a single value of V_r. We shall see in the next section how deviations from strictly plug flow can lead to both instability and alternative steady states.

Even with plug flow, however, another factor may be important in practice, and may produce control difficulties rather similar to those of instability. This is called 'parametric sensitivity', which may be defined as the situation where a small change in the operating variables leads to a very large change in the behaviour of the reactor. The steady state of the reactor is stable in the sense already referred to (i.e. perturbations occurring within the reactor die out, and the sequence of possible steady states is continuous) but the reactor may still be very difficult to control if a small change in an operating variable causes a large change in the (stable) steady state.

A case discussed is the behaviour of a simple exothermic reaction taking place in a tubular reactor whose walls are maintained at some value T_w which is the same over the whole length of the reactor [**4**]. The temperature T_r of the reacting fluid is supposed for simplicity to be uniform over the cross-section (i.e. there is assumed to be a temperature discontinuity $T_r - T_w$ at the wall) but it varies over the length on account of the heat of reaction. Solving the differential equations for this model of a reactor [**4**], the authors have shown that if T_r is plotted as a function of time along the reactor, curves are obtained as shown in Fig. 57 for various values of T_w.

It will be seen that quite a small increase in T_w causes the temperature in a certain region of the reactor to increase very sharply. That is to say a 'hot spot' develops and this shows itself also in a corresponding sharp fall in the reagent concentration (i.e. a sharp increase in the degree of conversion) as shown in Fig. 58. The reactor thus displays parametric sensitivity in relation to the value of T_w; a change in this variable of as little as 1 K has the effect, in one of the examples studied, of reducing the volume of reactor required for virtually complete conversion by more than a half. A similar sensitivity occurs in relation to the heat-transfer coefficient and also to the degree of dilution of the reaction mixture. The initial temperature of the packing material can also be important [**14**].

Such behaviour is easily understandable on intuitive grounds. Thus if T_w is held at a value sufficiently below the inlet temperature of the reacting fluid, the latter will cool off and reaction will become very slow. However, as T_w is raised above a value which is more or less critical, the combined effects of reaction heat and the exponential dependence of reaction rate on temperature cause the fluid temperature to rise steeply

Fig. 57. Parametric sensitivity: influence of wall temperature on reaction temperature (after Bilous and Amundson, *A.I.Ch.E. Journal*, 1956, **2**, 117).

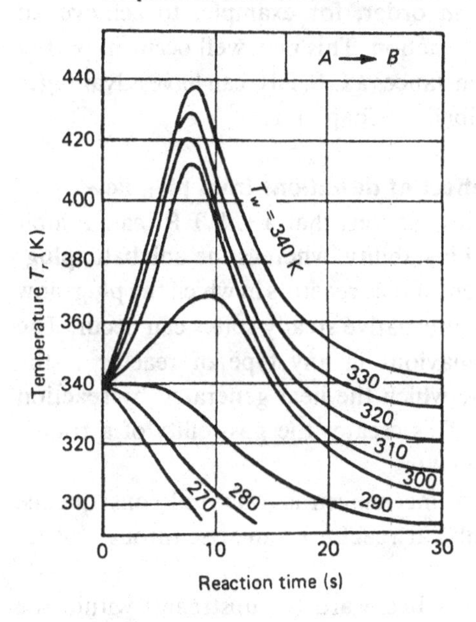

Fig. 58. Parametric sensitivity: influence of wall temperature on degree of conversion (after Bilous and Amundson, *A.I.Ch.E. Journal*, 1956, **2**, 117).

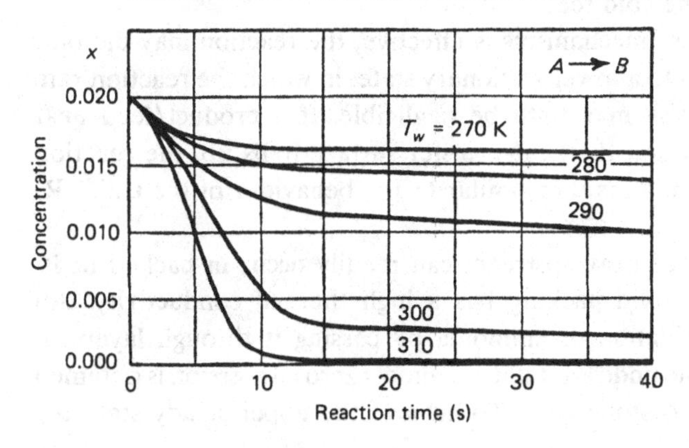

above its inlet value, this rise being only limited by the progressive depletion of the reagents.

Although this example was concerned with a simple reaction, the same considerations clearly apply to more complex types of reactions and for these the results have important industrial applications. For example, if the useful reaction were accompanied by a wasteful and exothermic degradation process, it will be evident from Fig. 57 that the onset of this process may occur rather quickly if the sensitive parameter exceeds a certain value. Nevertheless it may be necessary for this parameter to be kept close to its critical value in order, for example, to achieve an acceptable yield from the useful reaction. This may well occur in partial oxidation processes, and in such instances a C.S.T.R. can have advantages over a tubular reactor, as described in Chapter 1.

9.7 Tubular reactors; the effect of deviations from plug flow

We have seen in previous sections that a C.S.T.R. can exhibit alternative stationary states, and instability, whereas the adiabatic plug-flow reactor cannot. If we consider tubular reactors in which the plug-flow criterion has been relaxed, then alternative steady states can occur. The essential condition for such behaviour in any type of reactor is the existence of some mechanism by which the heat generated by reaction can pass to an earlier stage [**3**]. This creates the possibility of a transit between lower and upper stable states.

With a C.S.T.R., this feedback mechanism is a natural consequence of the mixing. In the case of a tubular reactor it can arise through either of the following circumstances:

(1) There is conduction of heat backward (or upstream) within the reactor.

(2) There is transfer of heat, e.g. with a heat exchanger, between the hot product and the cold feed.

If neither of these mechanisms is effective, the reaction may die out, i.e. the system falls to a lower stationary state, in which the reaction rate and temperature rise may both be negligible. If a product/feed heat exchanger is used, and the heat-transfer surface fouls up, the reaction may also die out, in a manner similar to the behaviour of the C.S.T.R. in Example 9.9.

The conduction of heat upstream can readily occur in packed beds, particularly if the solid packing has a high thermal conductivity. An example is the oxidation of ammonia by passing it through layers of platinum gauze. The undesired state of almost zero conversion is obtained when the catalyst is stone cold. To achieve the upper steady state the catalyst must initially be heated. This state will then maintain itself provided the catalyst retains its chemical activity and is not poisoned. In practice this is not difficult to ensure, but in other cases, e.g. the oxidation of isopropanol over a copper catalyst [**11**], extinction can easily occur if conditions are not closely controlled.

It is not essential to have a conducting solid present. In the case of flames the backward conduction of heat occurs in the fluid stream (though the backward diffusion of free radicals also plays a significant role). If this feedback is inadequate, the cold gas entering the reaction zone is not brought up to ignition temperature, and the flame 'blows out'.

Heat transfer between product and feed can be obtained in a heat exchanger external to the reactor, but it can also be advantageous to arrange for such heat exchange inside the reactor, between incoming feed, and the hotter, reacting, mixture. We shall consider this in more detail in a later section, but Fig. 59 shows diagrammatically what can occur when there is a counter-current heat exchanger within the reactor. Here the lower curve refers to the heating of the gas passing upwards through the heat exchanger tubes and the upper curve refers to the temperature changes taking place in the gas passing downwards through the catalyst. In the latter there is initially a temperature rise, due to the heat generation by reaction being faster than the heat exchange; lower down the reactor, where the reagents are becoming depleted, the heat exchange overtakes the heat generation and the gas temperature falls progressively to its outlet value. (The latter naturally exceeds the inlet

Fig. 59. Temperature distribution in tubular reactor having heat interchange with cold ingoing fluid.

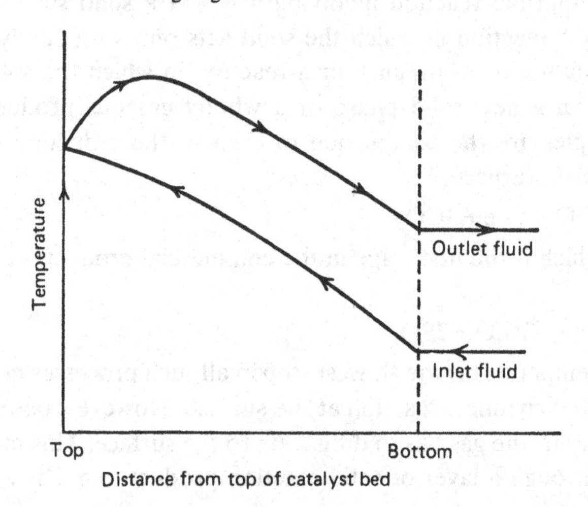

temperature and by an amount which, whenever there is zero heat loss from the system as a whole, depends only on the extent of reaction, on the heat of reaction, and on the heat capacities.)

From a calculation of these temperature profiles [**2, 3**], two important conclusions can be drawn: (1) for a given set of conditions there is a certain minimum value of the heat-transfer coefficient below which reaction will die out; (2) the effect of any progressive deterioration of the catalyst is that the heat exchange capacity which is necessary in order to keep reaction going must be gradually increased.

We shall see later that in a reaction such as ammonia synthesis the optimum sequence is one of diminishing temperature through the depth of the catalyst. It will be clear that this cannot be attained in the type of counter-current flow system which has been described above. The best that can be hoped for is an approximation to the optimum sequence in that part of the reactor which corresponds to the region to the right of the peak temperature shown in Fig. 59. The alternative is to make some radical change in the mode of operation of the reactor such as the introduction of 'cold shots', as will be described.

9.8 Diffusion control or kinetic control. Temperature effects

In Chapter 7 we considered the behaviour of catalytic reactors, and we saw that mass-transfer effects could become rate-controlling when the catalyst surface was very active. In exothermic heterogeneous reactions the combination of diffusional and kinetic processes can give rise to a sigmoid-shaped heat generation curve, and thereby to upper and lower stable reaction states.

Consider an exothermic reaction involving a gas at a solid surface. This may be either a reaction in which the solid acts only as a catalyst (e.g. ammonia oxidation on platinum), or a reaction in which the solid is consumed to form a new solid phase or a wholly gaseous product. Well-known examples are the combustion of carbon, the reduction of iron oxide in the blast furnace,

$$Fe_2O_3 + 3CO = 2Fe + 3CO_2,$$

and the reaction which is the first stage in the commercial production of zinc,

$$2ZnS + 3O_2 = 2ZnO + 2SO_2.$$

At low enough temperatures the slowest step in all such processes may be expected to be the chemical reaction at the surface. However, before this reaction can occur, the gas has to diffuse up to the surface. This may involve diffusion through a layer of solid reaction product (e.g. ZnO in the above example) or through the other gases present in the system. In all instances a diffusional process must precede chemical reaction – and a further diffusional process, namely the outward diffusion of gaseous reaction product, must also occur subsequent to reaction in all instances where such products are formed. Since the temperature coefficient of diffusion is normally much smaller than that of chemical reaction, the diffusional processes usually become intrinsically slower than the surface reaction at a high enough temperature and this gives rise to a diffusion-controlled regime. The result is a sigmoid-shaped curve for the rate of heat generation, although of a rather different shape from the case considered in § 9.2.

In Chapter 2 we discussed the case of a first-order reaction in which mass transfer resistance was also a factor. Equation (2.41) can be rewritten as

$$\text{Rate} = \frac{(kD/\delta)c_b}{k + D/\delta}. \tag{9.9}$$

Here k is the first-order velocity constant, D the appropriate diffusion coefficient, δ the 'stagnant-film thickness', and c_b the bulk concentration of reagent. We saw in Chapter 2 that at low temperatures the rate is likely to be kinetically controlled ($k \ll D/\delta$), whereas at higher temperatures the rapid increase of k with temperature can lead to diffusional control ($k \gg D/\delta$).

In Equation (9.9) let us put

$$k = \text{const.}\ e^{-E_A/RT} \quad \text{and} \quad D = \text{const.} \times T^2, \tag{9.10}$$

which expressions describe the characteristic temperature-dependence of k and D. Assuming that we know what the constants are in equation (9.10), and also assuming that δ in equation (9.9) does not vary significantly with temperature, we can plot the rate as a function of temperature. The result is indicated in Fig. 60. This shows how the rapid increase in rate which is characteristic of chemical reaction gives way to the slower increase corresponding to the onset of diffusion control.

We saw earlier that with an exothermic reaction the depletion of reagent caused a sigmoid heat generation curve. The corresponding rise of temperature during the course of the reaction gives rise to the possibility of transition from chemical rate control to diffusional control, and we see from Fig. 60 that this too can lead to a sigmoid shape of curve for the reaction rate or rate of heat generation. The two effects will reinforce each other. The reacting solid will tend to heat up as the rate of its reaction with the gas increases, and it is the flow of gas which takes away the heat of reaction. The analogy with Fig. 49 is not exact, for that refers to a C.S.T.R. in the steady state, but the behaviour is qualitatively similar. If the gas entering the reactor is not hot enough the reactor will not, of itself, ignite to a high conversion state with a high temperature of effluent gas. A coal stove, with cold air blown through it, will not light. If the coal is heated (electrically, or by the combustion of an initial quantity of more inflammable fuel – such as a fire-lighter) then the reaction may become self-sustaining. Alternatively if the air is heated (as in an electrically-heated blower) the coal may be heated to a temperature causing ignition. The reaction will subsequently proceed with cold feed air and the heated blower can be switched off.

If the solid is a catalyst then a steady state of high conversion can be set up, in which the rate is diffusion controlled. If the solid is a reagent, and is itself consumed, then the reactor will show behaviour varying with time, perhaps slowly. A coal stove will burn out if new coal is not provided. The reaction zone will move through the bed until all the fuel has been consumed. The next section will consider this in a little more detail.

9.9 The propagation of reaction zones

As we have just seen, interesting problems can arise where a reaction zone passes through a bed of solid particles which are reacting

Fig. 60. Transition from chemical control to diffusion control. The full curve is the plot of equation (9.9). The two dotted curves correspond to the influence of temperature on (*a*) diffusion unlimited by reaction, and (*b*) reaction unlimited by diffusion.

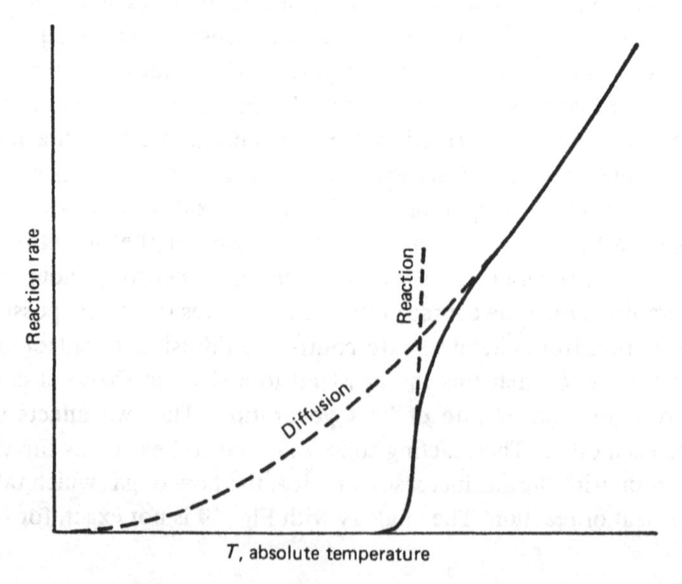

with a gas flowing through the bed. A hot zone of reaction will tend to move in both directions. It will move upstream against the flow of cool feed gas, and here conduction in both phases will play a part. It will also tend to move downstream both by such conduction and also because the hot product carries heat by convection. In the latter case, however, if the gas reagent is fully converted (e.g. there is no oxygen left from the air entering the stove) the bed will be heated downstream, but the solid there will not react. As the reacting solid upstream is consumed, the gas arriving downstream will not be denuded of reagent, and the heated solid there will start to react. The reaction zone will then move downstream.

As well as the coal stove, we have examples of this behaviour in the roasting of sulphide ores. A typical reaction scheme is

$$2ZnS + 3O_2 \rightarrow 2ZnO + 2SO_2.$$

In a blast furnace we have a similar situation, but with a more complicated set of reactions. In this case the liquid metal is removed from below the reaction zone, and new ore/fuel mixture is fed to the reaction zone from above. Thus the reaction zone is maintained stationary within the blast furnace.

It frequently occurs with organic reactions which are heterogeneously catalysed that 'coke' is laid down on the catalyst, due to (often undesired) cracking reactions of the feed hydrocarbon. Such carbonaceous material can seriously reduce the reactivity of the catalyst, and it has to be removed. This 'regeneration' is usually carried out by burning it off in a current of hot air. Since the reaction is highly exothermic, a reaction zone passes through the bed in the same direction as the air flow, as shown in Fig. 61. An important factor in the operation of this process is that the peak

Fig. 61. Diagrammatic representation of the propagation of a reaction zone.

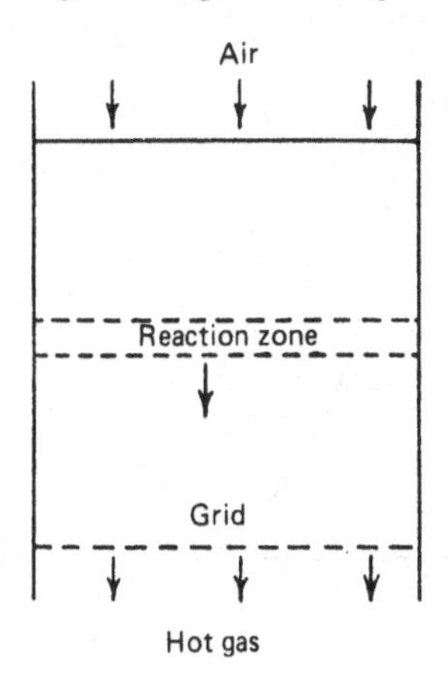

temperature of the catalyst should not exceed a value at which the catalyst changes structure, perhaps by sintering, and loses its activity. For this to be avoided it is common for the reaction to be started with very 'lean' mixtures of air and nitrogen, the concentration of oxygen being gradually raised as the carbon on the bed is reduced.

In such systems the problem is to calculate the temperature profiles of solid and gas, and the rates at which these profiles move through the bed. Fig. 62 gives a typical pair of profiles at a given time. The problem is rather a difficult one, which demands computer calculations. The calculations are sensitive to the assumed model, and to the parameters chosen for the model. In fact, systems such as these can show extreme parametric sensitivity (see § 9.6).

An early example of such a calculation [**15**] is indicated in Fig. 63. The peak temperature may rise as the front moves down the bed. It can

Fig. 62. Temperature profiles at a particular instant.

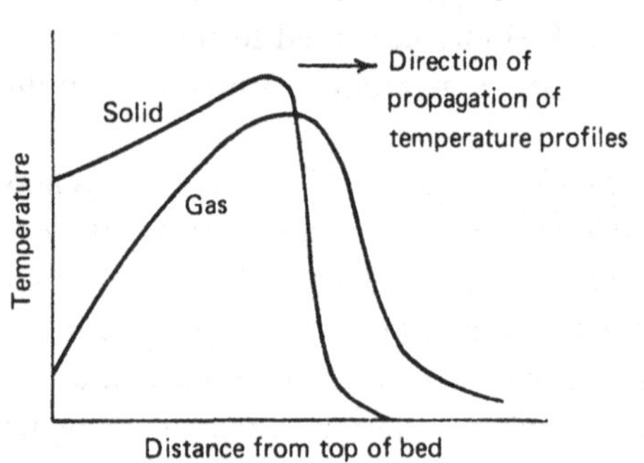

Fig. 63. Temperature profiles at two instants (based on Johnson, Froment and Watson [**15**]).

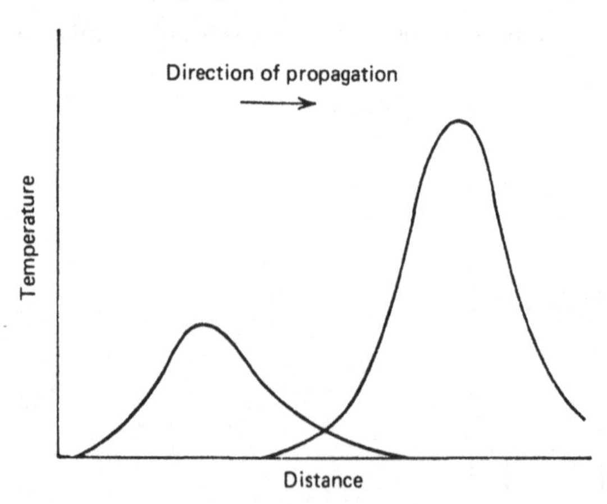

rise to very high values indeed if the reaction front moves at the same speed as that at which a pulse of heat would be carried through the bed. In such a case the reaction heat is 'trapped' within the moving reaction front, and severe damage can result. We shall not discuss this matter further here; there has been much work on this problem described in the literature since [**15**].

Example 9.15

Carbon is being burnt off a coked catalyst. If the reaction front dividing the unburned coke and the clean catalyst is a sharp one at which the reaction takes place rapidly, show that this front moves through the bed with a speed given approximately by

$$\text{rate of advance} = \frac{12 g y_0}{\rho x_c}.$$

In this example g = molar flow rate of gas (kmol s^{-1} per m^2 of bed cross-section), y_0 = mole fraction of oxygen in the gas fed to the bed, ρ = kg of catalyst per m^3 of bed volume, and x_c = weight fraction of carbon on coked catalyst.

Example 9.16

Show that, given infinitely rapid heat exchange between the gas phase and the solid packing, a sharp temperature front will move through a packed bed with a speed given approximately by

$$\text{rate of advance} = \frac{g C_g}{\rho C_s}.$$

Here g and ρ are as in Example 9.15, C_g is the *molar* specific heat of the gas (J kmol^{-1} K^{-1}) and C_s is the specific heat of the packing (J kg^{-1} K^{-1}).

Note that this front could be produced by the heat of reaction, e.g. in a regeneration of coked catalyst. If it were to move at the same speed as that in Example 9.15 a serious rise in temperature (in principle increasing with time without bound) would occur at the front as it moved down the bed.

Symbols

A Heat transfer area, m^2.
c Concentration, kmol m^{-3}.
C Specific heat, J kg^{-1} K^{-1}.
D Diffusion coefficient, m^2 s^{-1}.
E Energy of activation, J kmol^{-1}.

G Feed flow rate, kmol s^{-1}.
ΔH Enthalpy change in reaction, J $kmol^{-1}$.
k Velocity constant.
Q Heat generation, or removal, J s^{-1}.
r Reaction rate, kmol m^{-3} s^{-1}.
T Temperature, K.
U Heat-transfer coefficient, J m^{-2} K^{-1} s^{-1}.
v Flow rate, m^3 s^{-1}.
y Conversion.
Z Frequency factor in rate constant.
δ Film thickness, m.
ρ Density, kg m^{-3}.

References

1. Pismen, L. M., *Chem. Engng Sci.*, 1980, **35**, 1950.
2. van Heerden, C., *Ind. Eng. Chem.*, 1953, **45**, 1242.
3. van Heerden, C., *Chem. Engng Sci.*, 1958, **8**, 133.
4. Bilous, O., and Amundson, N. R., *A.I.Ch.E. Journal*, 1955, **1**, 513.
5. Hoftijzer, P. J. and Zwietering, T. N., *Chem. Engng Sci.*, 1961, **14**, 241.
6. Westerterp, K. R., *Chem. Engng Sci.*, 1962, **17**, 423.
7. Aris, R., *Elementary Chemical Reactor Analysis*, (Prentice-Hall, 1969).
8. Perlmutter, D. D., *Stability of Chemical Reactors*, (Prentice-Hall, 1972).
9. Schmitz, R. A. and Amundson, N. R., *Chem. Engng Sci.*, 1963, **18**, 265.
10. Heemskerk, A. H., Dammers, W. R. and Fortuin, J. M. H., *Chem. Engng Sci.*, 1980, **35**, 439.
11. Frank-Kamenetskii, D. A., *Diffusion and Heat Exchange in Chemical Kinetics* (Princeton University Press, 1955). Based on the Russian edition of 1947.
12. Dutton, J., Ph.D. Thesis, London University, March 1968.
13. Liao, P. C. and Wolf, E. E., *Chem. Engng Comm.*, 1982, **13**, 315.
14. Liu, S.-L., Aris, R. and Amundson, N. R., *Ind. Eng. Chem. Fund.*, 1963, **2**, 12.
15. Johnson, B. M., Froment, G. F. and Watson, C. C., *Chem. Engng Sci.*, 1962, **17**, 835.
16. Gray, P. and Scott, S. K., *Chem. Engng Sci.*, 1983, **38**, 29.

10

Optimization: concluding comments

10.1 Introduction

In earlier chapters we have touched upon the question of the stability and controllability of chemical reactors. In this, the final chapter, we wish to introduce the reader to some elementary considerations of reactor optimization.

'Optimization' in chemical reaction engineering has come to mean the application of certain mathematical techniques for the purpose of discovering those values of the design, or operating variables, which are in some sense 'the best'. The criteria of judgement may be rather narrow, or the techniques somewhat simplified, but the approach has often led to substantial improvements of design or operation.

What is usually aimed at in optimization studies is the maximization of the profit of a process, or the minimization of its costs. As was remarked earlier in § 1.2, this is very imperfect; the conventional definitions of profit and cost are far from representing the total influence of an industrial process on the community; much that is of importance gets left out in this kind of analysis.

However, those who do optimization studies often get criticized from the opposite standpoint. It is said that, far from being too much 'cost conscious', they are not sufficiently so; instead of trying to optimize a real process in terms of cash, they are content to set up artificial models of the process and to use some non-monetary variable, e.g. reaction yield, which may have only a partial bearing on the economics.

Both forms of criticism obviously derive from the same basic difficulty. In seeking to optimize we have to confine ourselves to whatever can be expressed quantitatively and this usually requires a considerable simplification of the actual situation. In other words the *objective function* whose maximum or minimum we can readily obtain is not the one we should really like to study, but is the best approximation to it that we can manage.

This is as true of reactor optimization as of any other sort. The kind of optimization we can achieve concerns the discovery of the optimum of a simplified objective function (e.g. yield, or output, or some part of the total costs) whose usefulness depends on its being not too far displaced from the optimum of the objective function we should be really interested in, if it were calculable.

Nevertheless, it is believed that these simplified studies can be of real value and especially, perhaps, in developing the attitude of mind which, when confronted by a particular process, is able to detect intuitively *that an optimum will exist.* The important thing is to use the results in a heuristic manner and without being tempted to believe that the calculated position of the optimum as obtained from the simplified objective function is necessarily its 'best' position.

10.2 Output and yield problems

We have already considered certain problems of this type. In § 4.3 we noted that when a second-order reaction is carried out in a chain of isothermal stirred tanks there is a certain ratio of tank sizes which minimizes their total volume, i.e. this ratio maximizes the *output* of the reactor system per unit total volume. In Chapter 6 we saw how to maximize the fraction of reagent converted to useful product (in certain cases of isothermal, competing, reactions). This procedure led to maximizing the *yield.*

These two examples illustrate two useful categories of optimization [**1**], using simplified objective functions:

(1) *Output problems.* These are concerned with the attainment of the maximum of output, i.e. of the amount of reaction product per unit of time and of reactor volume.

(2) *Yield problems.* In these the problem is to maximize reaction yield, i.e. the fraction of reagent converted to desired product. This is done by appropriate choice of the conditions which suppress the formation of substances which are not desired.

The significant difference between the two classes is that in the first the minimization of reaction time is the essence of the problem whereas in the second it is of no significance. In the latter it is assumed that whatever contact time is needed to maximize the yield can be afforded.

Therefore output problems will have the greater economic significance in situations where there are no side reactions or where the capital cost associated with large reactor volume far outweighs the daily costs of the reagents. Yield problems, on the other hand, tend to become more significant in complex organic reactions; here the yield may often be much less than 100% and, if reagent costs are high relative to other costs, a small gain in yield may be extremely valuable.

The occurrence of the two classes of problem will first be reviewed in a physico-chemical and intuitive manner. It will be argued that many types of reaction are favoured by using an *optimum temperature sequence.* The way in which this may be calculated for a particular type of reaction will be described in a later section.

The instance of *reversible exothermic reactions* is one of the best studied of the output problems. These reactions include many which are important industrially, e.g. ammonia and methanol synthesis, the oxidation of sulphur dioxide, the hydration of ethylene:

$$C_2H_4 + H_2O = C_2H_5OH,$$

and the water-gas shift reaction:

$$H_2O + CO = H_2 + CO_2.$$

For such reactions the effect of a temperature increase is to raise the speed of the forward reaction but also to lower the value of the equilibrium constant and thereby to reduce the maximum attainable conversion. Thus the kinetic and thermodynamic factors are, in a certain sense, in opposition.

Such reactions are commonly carried out in tubular reactors. It follows from what has been said that in the entrance region, where the reacting gas is still far from equilibrium, it is advantageous to use a fairly high temperature in order to increase the reaction rate. In the outlet region however the temperature should be brought to a much lower level so that a better equilibrium state may be approached. In fact at each cross-section there is a certain optimum temperature such that, if these temperatures could be achieved along the whole length, their effect would be to maximize the output of the vessel. This problem has been discussed by many authors, e.g. [**2**–**4**].

Another important example of an output problem occurs in the system of consecutive reactions:

$$A + B \xrightarrow{1} X \xrightarrow{2} Y,$$

where it is X which is the desired product. This system has been discussed already, in § 6.4, from the standpoint of optimum time, but it will now be considered from the standpoint of optimum temperature. Let it be supposed that $E_2 > E_1$. In this situation the temperature should be high initially, in order to accelerate the first reaction (and thus to obtain a large output from a given reactor volume), but the temperature should be reduced progressively, as X accumulates, in order to utilize the fact that a reduction of temperature slows down the wasteful reaction, $X \rightarrow Y$, more than the useful reaction $A \rightarrow X$. Therefore there is again an optimum temperature sequence which diminishes continuously from the reactor inlet [**2, 4**].

A further example which provides a bridge between the output and yield problems should also be mentioned. This concerns the parallel reactions

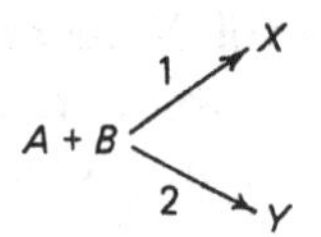

where X is the desired product and $E_2 > E_1$. Considering the matter first from the standpoint of maximizing the output of X, there is here an optimum sequence of temperatures. It is one in which the temperatures should *increase* from inlet to outlet. The reason for this is less intuitive than in the previous examples. In the region of the reactor inlet it is best to maintain a low temperature in order to favour conversion to X rather than to Y; in the region of the outlet a higher temperature becomes preferable in order to raise the otherwise rapidly diminishing reaction rate – by this means more X is gained, even though it also results in more Y. On the other hand, if time is of no significance, i.e. in the corresponding yield problem, there is no optimum sequence; the temperature should obviously be as low as possible throughout the length of the reactor. The distinction between the two types of problem is clearly indicated.

Example 10.1

A reaction $A \rightarrow B$ giving the desired product B, is accompanied by side reaction $A \rightarrow C$ giving the undesired substance C. The activation energy of the second reaction is known to be greater than that of the first.

It is proposed to carry out the reaction process for a fixed time t_e and at a *constant* temperature T, which may however be chosen at any required value. Why is there an optimum value of T which results in maximum conversion of A to the desired product? Using suitable simplifying assumptions (e.g. concerning the kinetic orders of the two reactions) obtain an equation from which the optimum value of T may be calculated.

In what respect would the nature of the problem of interest, and the conclusion to be drawn, be quite different if no limit were imposed on the duration of reaction?

What has been said above about optimum temperature sequences in a tubular reactor applies, *mutatis mutandis*, if the above reactions were carried out in a chain of stirred tanks. In this case there would be a series of optimum stationary temperatures for the various tanks. However, a further set of variables is now available for optimization and these are the relative volumes or the holding times.

Turning now to some examples of yield problems, the reader may first be reminded that the *size* of the reactor is not here regarded as being of any consequence. If a certain change in the operating conditions results in an increase of yield, this is regarded as a gain, even though such a change might also require a larger reactor to attain the same daily output.

The following degradative reaction scheme requires an optimum temperature as well as an optimum time* for maximum yield [**5**].

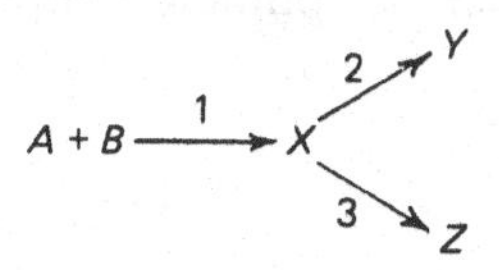

Here X is the desired product. If it occurs that $E_2 < E_1 < E_3$, and if it may be assumed for simplicity that all reactions are first order, the yield of X, relative to the amount of reagent entering the vessel, is highest when reaction is carried out at a *particular temperature.* Somewhere between the limits of the low temperature (which promotes $X \rightarrow Y$ more than $A \rightarrow X$) and the high temperature (which promotes $X \rightarrow Z$ more than $A \rightarrow X$) there is this special temperature level which minimizes the sum of the two wasteful reactions relative to the useful one. Moreover, when all three reactions are of the same order, the temperature level in question is independent of the concentrations and it is for this reason that it is a single temperature and not an optimum sequence.

Example 10.2

A reagent mixture decomposes by three simultaneous first-order irreversible reactions, giving products X_1, X_2, X_3. The energies of activation are in the order $E_2 < E_1 < E_3$. If X_1 is the desired product,

* The simpler scheme

$$A + B \xrightarrow{1} X \xrightarrow{2} Y$$

requires an optimum time (as shown in § 6.4) but not an optimum temperature when the matter is regarded solely from the yield standpoint. If X is the desired product and $E_1 < E_2$ the temperature should be as low as possible. Conversely if $E_1 > E_2$ it should be as high as possible.

show that the optimum temperature is given by:

$$T_{opt} = (E_3 - E_2)/R \ln\left\{\frac{Z_3(E_3 - E_1)}{Z_2(E_1 - E_2)}\right\}.$$

Discuss the optimum temperature sequence if the reactions have different orders.

What is probably a more commonly occurring situation in the production of organic chemicals is where the required product is formed via one or more intermediates. The latter may be unstable and short-lived substances such as free radicals, or they may be substances which are quite stable (thus monochlorbenzene is an intermediate in the formation of dichlorbenzene from benzene). If it occurs that the original reagents and also the intermediates can undergo side reactions, the situation is one which offers important scope for the use of optimum temperature sequences [**1**].

Let the process be represented:

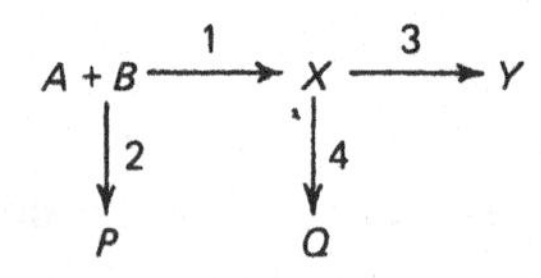

where here it is Y which is the desired product, whilst P and Q are waste products. The yield of Y is obviously higher the greater is the speed of reaction 1 relative to reaction 2 and of reaction 3 relative to reaction 4.

Let $E_1, \ldots, E_4$ be the activation energies and consider first the case $E_1 < E_2$ and $E_3 > E_4$. As far as the competition of reactions 1 and 2 is concerned a low temperature is preferable, since E_2 being greater than E_1 means that a reduction of temperature depresses the rate of reaction 2 more than it depresses the rate of reaction 1. Conversely, as regards the competition of reactions 3 and 4, a high temperature is desirable since this has the effect of favouring conversion of X into Y instead of into Q. However, considering both factors together, the temperature should be fairly low during the initial stages of reaction whilst X is accumulating, but it should be much higher during the later stages when the main reactions are the conversion of X into Y or into Q. However, since all four reactions actually take place simultaneously, although to varying extents, the temperature should be increased progressively. Thus, if reaction were carried out in a tubular reactor there would be an optimum temperature sequence for highest yield. If it were carried out in a chain of stirred tanks there would be a certain optimum temperature for each vessel in the cascade.

Similar considerations apply to the converse case, $E_1 > E_2$, $E_3 < E_4$, except that the optimum sequence is now a diminishing one*. Using typical values for the activation energies quite remarkable increases of yield can be obtained. Thus, in an example where the activation energies were chosen as differing by 24 000 J $kmol^{-1}$, the highest yield obtainable by *isothermal operation* was calculated to be 25%, and this required a temperature of 53 °C, which is the best possible *single* temperature level. By carrying out the same reaction in two tanks in series, one held at 7 °C and the other at 141 °C, the yield could be raised to 53%. By using more tanks in series or a tubular reactor, the yield could be brought to well over 60%, i.e. more than double the value obtainable by isothermal operation. The theory of this effect has been extended by several authors [**4–7**].

So much for a brief review of several of the reaction schemes for which optimum temperatures and optimum temperature sequences have been studied. For details of how the optima may be calculated the reader is referred to the literature. There are also reactions in which it would be desirable to vary the pressure (or concentration) during the course of reaction; this leads to an optimum pressure sequence [**8**].

Detailed calculations of this sort require accurate information about the activation energies and kinetic constants. When this is not available, a knowledge of the qualitative behaviour of the system, based on hit-and-miss trials in the laboratory of rising or falling temperature sequences, can still be very useful.

10.3 The reversible exothermic reaction

The reason why this type of reaction requires a falling temperature sequence has already been outlined, and the means of calculating the sequence will now be described. It is the simplest of the non-isothermal optimization problems in regard to its mathematical demands.

In the earlier discussion the matter was presented in terms of an opposition between the kinetic and thermodynamic requirements of the reaction. However, since equilibrium is normally not attained under industrial conditions, it is preferable to carry through the discussion in kinetic terms entirely. This may be done by reference to the fact that, for an exothermic reaction, $E < E'$ where E and E' refer to the activation energies for the forward and backward reactions respectively. This inequality, which is discussed in § 2.5, means that the effect of raising the temperature is to increase the speed of the reverse reaction more

* The other two possibilities, $E_1 > E_2$, $E_3 > E_4$ and $E_1 < E_2$, $E_3 < E_4$, do not require optimum sequences but rather uniformly high and uniformly low temperatures respectively.

than that of the forward one. Bearing in mind that we are here concerned with an output problem, it follows that in parts of the reactor close to the inlet, where little reaction product has yet been formed, the temperature should be high in order to accelerate the forward reaction and thereby to raise the output per unit of reactor volume; however, as reaction product accumulates the temperature should be progressively reduced, since this has the effect of lowering the speed of the backward reaction more than that of the forward one. Therefore, as was said previously, the optimum sequence is one of diminishing temperatures from inlet to outlet.

The question to be answered is this: what should be the temperature at any cross-section of a tubular reactor which will result in a *minimum* volume of reactor (or minimum quantity of catalyst) for the attainment of some chosen hourly output of product?*

Intuitively the necessary condition would appear to be that the *net speed of reaction* (i.e. the forward rate minus the backward rate) *must have its highest possible value at every cross-section.* This condition can be proved, as follows.

Consider equation (3.3):

$$V_r = G\int_0^{y_e} \frac{dy}{r} \quad \text{if } y_i = 0; \tag{10.1}$$

G is the mass flow rate and V_r is the reactor volume necessary to achieve a composition y_e of product at the reactor outlet. r is the reaction rate; for a given inlet gas composition and given operating pressure,† this is a function only of the degree of conversion and of the temperature. Thus

$$r = f(y, T).$$

Let T_m be the value of T which makes r have a maximum value r_m for a particular value of y. Then for any other value of T and the same value of y we have

$$\frac{1}{r} > \frac{1}{r_m} \tag{10.2}$$

Hence

$$\int_0^{y_e} \frac{dy}{r} > \int_0^{y_e} \frac{dy}{r_m}. \tag{10.3}$$

This must be true because if the *integrand* $1/r$ is greater than the integrand $1/r_m$ at the same value of y, the same inequality must also apply to the

* It can be shown [2] that this is equivalent to asking what is the temperature sequence which will *maximise* the hourly output for a chosen value of the reactor volume.

† Pressure drop along the reactor is here regarded as negligible. It may be noted that plug flow is also being assumed.

corresponding *integrals* when they are compared over an equal range of y values. Comparison of equations (10.1) and (10.3) now shows that all other possible values of V_r will be greater than the value obtained by having the maximum reaction rate r_m at all cross-sections of the reactor. In brief, the reactor is of minimum size if T is chosen so as to maximize the local reaction rate everywhere in the system.

For simplicity we shall discuss a gas reaction having the stoichiometry

$$A+B=C+D,$$

and it will be further assumed that the reaction rate is expressible in the form

$$r=kab-k'cd,$$

where $a, \ldots, d$, are concentrations. Using equation (2.2) the last equation may be re-written

$$r=Z\,e^{-E/RT}ab-Z'\,e^{E'/RT}cd. \tag{10.4}$$

If suitable values of Z, Z', E and E' (with $E'>E$) are substituted in this equation, r can be plotted as a function of T for fixed values of $a, \ldots, d$. The results are as shown in Fig. 64. Here the three curves refer to three sets of fixed concentrations, as derived from the same inlet gas, with degree of conversion increasing from curves 1 to 3.

Considering any one curve it will be seen that the rate first rises with increasing temperature, reaches a maximum and then falls rapidly. It becomes zero at the temperature T_e at which there would be thermodynamic equilibrium for the particular values of the concentrations. For all values of T greater than T_e the rate is negative, due to the second term on the r.h.s. of equation (10.4) having become larger than the first.

It has been shown already that the reactor volume will be a minimum of r is everywhere at its highest possible value. This corresponds to a process of moving from right to left down the dotted locus of the maxima as shown in Fig. 64. The position of this locus may be determined by differentiating equation (10.4) with respect to T and putting $(\partial r/\partial T)_{a,b,c,d}$ equal to zero.

However, as applied to the conditions of operation of a tubular reactor, this procedure will not be quite accurate. It is not the maximum rate at given values of the volume concentrations which is required, but rather the maximum rate at given values of the total pressure and of the degree of conversion.

It will be supposed that A and B enter the reactor in equimolar amounts and also that there is plug flow. If z is the degree of conversion at a particular cross-section, the corresponding mole fractions are: for A and B, $\frac{1}{2}(1-z)$; for C and D, $\frac{1}{2}z$. Assuming that the gas mixture is ideal, the concentrations at the cross-section are therefore

$$\left.\begin{aligned} a = b &= (1-z)P/2RT, \\ c = d &= zP/2RT, \end{aligned}\right\} \tag{10.5}$$

where P and T are the pressure and temperature respectively. Substituting these expressions in equation (10.4) we obtain

$$r = \left(\frac{P}{2RT}\right)^2 (Z e^{-E/RT}(1-z)^2 - Z' e^{-E'/RT} z^2). \tag{10.6}$$

If this is differentiated with respect to T and the differential coefficient is set equal to zero, the condition of maximum rate is obtained:

$$Z e^{-E/RT_m}(1-z)^2(E-2RT_m) - Z' e^{-E'/RT_m} z^2(E'-2RT_m) = 0,$$

where T_m is the optimum temperature. We have assumed, as on other occasions, that the temperature dependence of Z and Z' can be disregarded in comparison with the exp $(-E/RT)$ terms. This expression may be conveniently rearranged to give

$$\frac{Z e^{-E/RT_m}(E-2RT_m)}{Z' e^{E'/RT_m}(E'-2RT_m)} = \frac{z^2}{(1-z)^2}. \tag{10.7}$$

A trial-and-error solution of this equation therefore allows the optimum temperature to be worked out for any value of the degree of conversion, provided the kinetic quantities Z, Z', E and E' are known for the particular reaction.

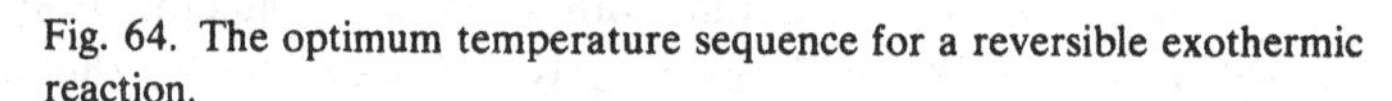

Fig. 64. The optimum temperature sequence for a reversible exothermic reaction.

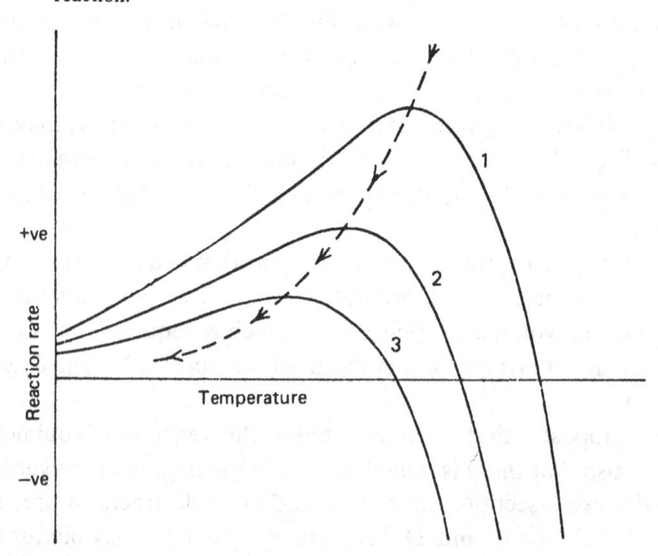

***Example* 10.3**

A reagent A undergoes a reversible gas-phase decomposition according to the equation

$A \rightleftharpoons B.$

Both the forward and backward steps are first order. The equilibrium constant for the reactions is 100 at 300 K and the decomposition of one kmol of A liberates 2.5×10^7 J. The velocity constant of the forward reaction is given by

$k = 10^9 \, e^{-10^4/T} \, s^{-1}.$

Estimate and plot the curve of reaction rate against temperature for a mixture corresponding to 50% conversion of A and at total molar concentration of 0.2 kmol m^{-3}. At what temperature is the net rate $A \rightarrow B$ a maximum for this composition? Compare your graph with Fig. 64.

[*Answer.* About 528 K.]

***Example* 10.4**

Consider the calculation of the previous example for the case $A \rightleftharpoons 2B$. Here the backward step is second order. The equilibrium constant for the reaction is 100 kmol m^{-3} at 300 K. The forward velocity constant and the heat of reaction are as in Example 10.3.

Again assuming a composition corresponding to 50% conversion of A, at what temperature is the reaction rate a maximum, if the total molar concentration at this composition is:

(a) 0.3 kmol m^{-3};
(b) 0.3 kmol m^{-3};
(c) 3 kmol m^{-3}.

[*Answers.* (a) 1224 K; (b) 631 K; (c) 425 K.]

It remains, however, to connect the optimum temperature with a particular position along the length of the reactor. For this purpose consider a volume element, dV, contained between two planes across the reactor normal to the flow direction. Let N be the molar flow rate of A (or B) at the reactor inlet. The degree of conversion being z and $z + dz$ at the two planes, the corresponding flow rates of C (or D) across these planes are zN and $(z + dz)N$ respectively.

Therefore by mass balance

$$(z + dz)N = zN + r\,dV,$$

or

$$V = N \int_0^z \frac{dz}{r}, \qquad (10.8)$$

where V is the volume of reactor from the inlet up to the cross-section where the degree of conversion has reached the value z. This equation is another form of equation (10.1).

The ready availability of computers enables the chemical engineer to carry out as an everyday matter optimization calculations which would have been impossible to contemplate, say, twenty years ago. This is not a book on computer techniques, so we will only outline here an approach to the problem of calculating the optimum temperature profile through the reactor. Our example, using equations (10.4)–(10.8), is both specific and simplified. More realistic cases can be handled in the same general manner.

We use the conversion, z, as our primary variable. For successively larger values of z:

(1) Calculate the optimum temperature, T_m, from a relationship such as equation (10.7).

(2) Given this value of T_m, calculate the maximum rate, r_m, by substituting T_m in equation (10.6).

(3) Integrate equation (10.8) numerically to find the volume V_{min}, which is the minimum volume to attain a given conversion, z.

We thus can obtain the optimum temperature profile, giving the maximum conversion, z, as a function of reactor volume V.

This idealized optimum has in practice to be modified; we shall consider only two such modifications here.

(*a*) For very low conversions, $z \to 0$, we see from equation (10.7) that $T_m = E/2R$. For typical values of E, this temperature may be far too high for any envisaged materials of construction. The computer program must therefore substitute for T_m the highest permissible temperature, T_p, and carry on. This is the same as saying that the reactor should operate *isothermally* at T_p, until T_m falls below T_p.

(*b*) There is an implicit assumption that we can, for $T_m < T_p$, actually control the temperature at T_m. We must therefore ask the computer to calculate the cooling duty, and the heat-transfer area, for this purpose. It may well turn out that it is not practicable to design a reactor which can adhere to this profile very precisely. It is no major problem for the computer program to incorporate restrictions of this practical type.

Returning to our example, we will assume for the purposes of illustration that the parameters have the following values, in SI units:

$$Z = 4.35 \times 10^{13}, \qquad E = 1.047 \times 10^{8},$$
$$Z' = 7.42 \times 10^{14}, \qquad E' = 1.255 \times 10^{8}.$$

A 70% degree of conversion at the outlet will be aimed at; the pressure will be taken as being 1 atm and the maximum permissible temperature, T_p, as 870 K. These various figures, which are taken from another example [**2**], result in equations (10.6) and (10.7) taking the forms

$$r_m = \frac{1.615\times 10^{15}}{T_m^2}\left(e^{-12590/T_m}(1-z)^2 - 17.06\, e^{-15100/T_m} z^2\right), \tag{10.9}$$

$$5.862\times 10^{-2}\, e^{2510/T_m}\left(\frac{6295-T_m}{7550-T_m}\right) = \frac{z^2}{(1-z)^2}. \tag{10.10}$$

Fig. 65 shows the temperature T_m and the degree of conversion plotted against the position in the reactor as measured by the variable V/N (the cumulative volume per unit feed rate of A or B). It turns out that the fraction of the reactor volume in which the temperature needs to be held at T_p is extremely small and is quite invisible in the figure. The region of the reactor where the reaction rate has become very small is that which naturally contributes by far the largest amount to the total volume;

Fig. 65. The optimum temperature sequence for a reversible exothermic reaction.

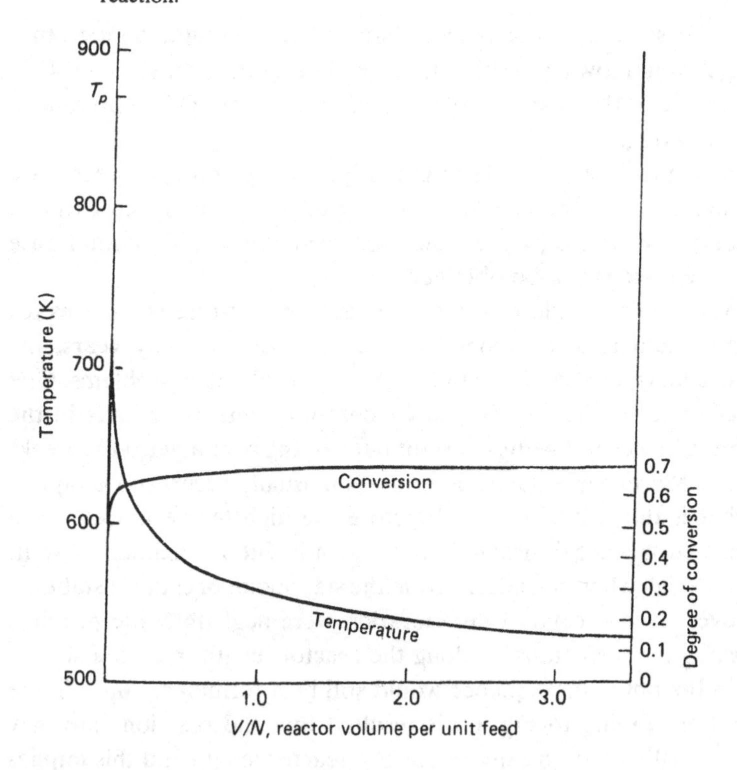

this may be seen from the figure to be the region in which the optimum temperature is not much above its outlet value of 530 K.

The result of the calculation is to show that, when the optimum temperature sequence is adopted, the necessary volume of the reactor, expressed as V/N, is 3.5 m^3 s $(kmol)^{-1}$. It is of interest to show that this is much smaller than would be required for isothermal operation. If equation (10.6) is substituted into equation (10.8), the latter can be integrated to give V/N as a function of the outlet degree of conversion together with the temperature, which is now taken as constant. If V/N is now minimized with respect to T it can be readily shown that the best possible isothermal temperature of operation is 550 K, and that the corresponding value of V/N is about 9.0.

***Example* 10.5**

With reference to the problem illustrated above, integrate equation (10.8) at constant temperature. Show thereby that operation at 550 K has the effect of minimizing the reactor volume for isothermal operation, as stated in the text. Show also that the necessary reactor volume, expressed as V/N, is about 9.0.

It will be seen that the reactor having the optimum temperature sequence, even allowing for the upper limit T_p, needs only about 40% of the volume of the best possible isothermal reactor. This represents a very large saving.

The saving can be particularly important in high-pressure reactions, such as ammonia synthesis, where the capital cost of the reactor can be high (see, for example [**9**]). Again, SO_2 oxidation [**3**] is another case where large savings can be obtained.

Although the principle of using optimum temperature sequences for reversible exothermic reactions has been known for many years, the practical achievement of the optimum poses considerable problems. One of the complicating factors is catalyst deterioration; this results in the optimum sequence not being constant but varying over a period of weeks or months. Moreover, catalyst deterioration usually occurs more rapidly the higher is the temperature. Therefore the high temperatures which are predicted as being desirable from the standpoint of maximum output, may be too high when considered from the standpoint of catalyst stability.

However, even if catalyst deterioration were negligible, the problem of adjusting the heat transfer along the reactor length in such a way as to obtain the optimum sequence would still be a formidable one. In the computation leading to Fig. 65 it resulted that the reaction rate may vary over a 100 000-fold range along the reactor length and this implies a corresponding variation in the rate at which the heat must be withdrawn from the catalyst. To achieve a *continuous* variation in the heat transfer through the wall, varying from inlet to outlet by this order of magnitude, is obviously almost impossible.

The perfect 'optimal reactor' is therefore not realizable. Nevertheless some approximation to the optimal sequence can be achieved in at least some regions of ammonia-synthesis plants, as Figs. 1 and 59 both show.

A *discontinuous* variation of heat transfer or temperature is, of course, much more easy to attain than a continuous variation. Accordingly, in many industrial reaction systems an approximation to the optimum sequence is obtained by stepwise change.

One such system is the multi-bed adiabatic reactor equipped with interstage cooling by means of heat exchangers. This is much used in SO_2 oxidation. The reactor consists of a series of catalyst chambers each of which operates adiabatically. Consequently in each of them the temperature *rises* from inlet to outlet and this, of course, is contrary to the theoretical ideal. However, by removing heat from the gas after it leaves each vessel, the temperature is brought to a much lower level at the entrance to the next vessel in the series. In brief the system is designed to approximate the optimum temperature sequence as a series of saw-tooth movements beneath the curve of the type of Fig. 65. This is illustrated in Fig. 66 where the axes are conversion and temperature rather than reactor volume and temperature. The larger the number of steps the closer the performance of the system approaches the theoretical optimum.

Fig. 66. Saw-tooth approximation to optimum temperature sequence.

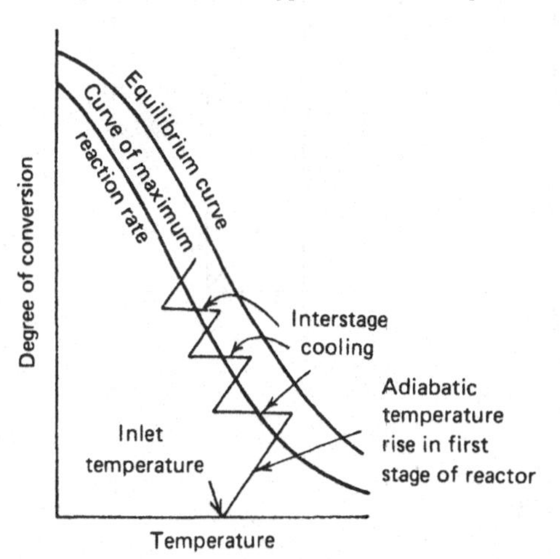

A second device is known as 'cold-shot' cooling. This again uses a number of adiabatic stages but here the method of cooling between stages is by the addition of an increment of cold reagents. That is to say, the total feed to the system is divided into a number of streams; one of them enters the first stage; a second mixes with the product of this stage and cools it before it enters the second stage; a third performs the same function at the third stage, and so on. Again there is a saw-tooth approximation to the optimum temperature sequence. This method is convenient in practice but for the same number of stages is cannot result in quite such a large output as can be obtained by use of direct cooling;

Fig. 67. 'Cold-shot' or 'quench' ammonia-synthesis reactor. (*a*) Gas inlet, (*b*) gas exit to heat recovery, (*c*) gas exit, (*d*) direct bypass, (*e*) gas from external start-up heater, (*f*) quench gas inlets, (*g*) pyrometer, (*h*) catalyst discharge nozzle. (By courtesy of Imperial Chemical Industries PLC.)

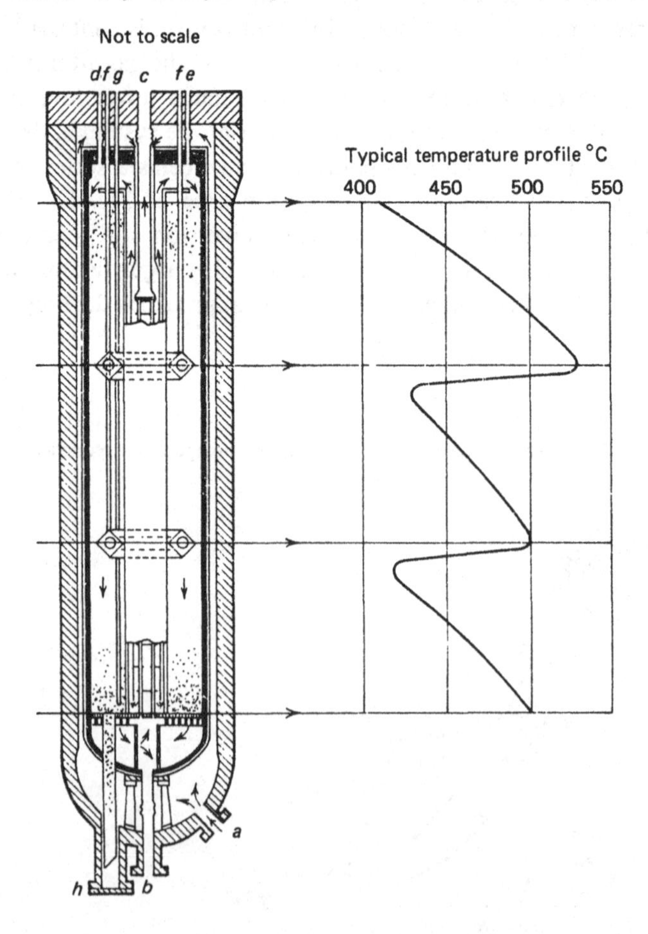

this is because the streams entering the second, and later, stages have a shorter mean residence-time in the system than if they had all entered the first stage in the normal way. A diagram of a 'cold-shot' ammonia converter is shown in Fig. 67.

10.4 Conclusion

The aim of this book has been to acquaint the reader with the simpler techniques of chemical reactor design. In recent years the literature on chemical reaction engineering has grown rapidly, together with (and fed by) the increasing availability and power of computers. The effective use of the computational power available nowadays requires a thorough understanding of the basic theory of chemical reaction engineering. Given that, the reader should be able to work out different solutions to a reactor problem, and hence select the best one for his particular situation.

However, the designer should be aware that optimization should also be carried out at other, successively higher, levels. The reactor is only a part of a chemical plant, and it is clear that a more complete optimization should take into account the necessary separation and purification equipment. This equipment may vary very widely according to the choice of reactor, and it is the cost of this equipment which may well determine which type of reactor is chosen, not the cost of the reactor alone.

Those concerned with managing a large concern must optimize at a still higher level. They must choose between different plants and processes. They must decide the scale of operation, and location, of individual plants, with the aim of optimizing the performance of the whole organization. Some extremely interesting applications of optimization techniques occur at this level, for example the use of linear programming to determine the optimum allocation of crude oil between refineries, and the optimum distribution of products to be made at those refineries.

Society itself is requiring, through governmental legislation, a greater say in determining the scale and manner of operation of different organizations in different fields. In recent years, public concern has been increasingly directed at such matters as industrial pollution and the effect of industrial activity on public health and safety. It is none too easy to apply the mathematical techniques of optimization at this level, since the problems themselves, and society's reactions to them, may be impossible to quantify precisely. If international considerations enter the picture, things get yet more difficult. Thus the problems of straight technology, such as this book has been concerned with, are in fact the easy ones!

***Example* 10.6**

A reversible gas-phase oxidation reaction, $X+O_2=2Y$, is carried out in a flow system, using air as the source of the oxygen. The gas leaving the reaction chamber, where the reaction comes close to equilibrium, passes to a cooler where the bulk of the product Y is condensed out as a liquid, at a temperature at which it has an appreciable vapour pressure. The remaining gas is discharged into the atmosphere.

Explain clearly why there may be an economic optimum with regard to the composition of the X–air mixture entering the reaction vessel. If the cost of X were the only significant factor affecting the economics of the process, show in outline how this optimum could be determined. How would the optimum composition be expected to vary with change of (*a*) the equilibrium constant of the reaction, (*b*) the vapour pressure of Y?

Symbols

a, b, c, d Reagent concentrations, see equation (10.4)
E Energy of activation, J, kmol^{-1}.
G Mass flow rate, kmol s^{-1}.
N Molar flow rate of reagent, kmol s^{-1}.
P Pressure, Pa.
r Reaction rate, kmol $\text{m}^{-3}\ \text{s}^{-1}$.
R The gas constant, J $\text{kmol}^{-1}\ \text{K}^{-1}$.
T Temperature, K.
V Reactor volume, m^3.
y Product composition.
z Degree of conversion.
Z Frequency factor in the velocity constant.

References

1. Denbigh, K. G., *Chem. Engng Sci.*, 1958, **8**, 125.
2. Bilous, O. and Amundson, N. R., *Chem. Engng Sci.*, 1956, **5**, 81, 115.
3. Boreskov, G. K. and Slinko, M. G., *Chem. Engng Sci.*, 1961, **14**, 259.
4. Aris, R., *Chem. Engng Sci.*, 1961, **13**, 18, 75, 197.
5. Horn, F., *Z. Elektrochem.*, 1961, **65**, 209.
6. Horn, F., *Chem. Engng Sci.*, 1961, **15**, 176.
7. Storey, C., *Chem. Engng Sci.*, 1962, **17**, 45.
8. van de Vusse, J. G. and Voetter, H., *Chem. Engng Sci.*, 1961, **14**, 90.
9. van Heerden, C., *Ind. Eng. Chem.*, 1953, **45**, 1242.

INDEX